THE SCIENTIST AND THE THEOLOGIAN

James P. Mackey

The Scientist and the Theologian

ON THE ORIGIN AND ENDS OF CREATION

the columba press

First published in 2007 by
the columba press
55A Spruce Avenue, Stillorgan Industrial Park,
Blackrock, Co Dublin

Cover by Bill Bolger
Origination by The Columba Press
Printed in Ireland by ColourBooks Ltd, Dublin

ISBN 978 1 85607 569 5

Table of Contents

For Seren Grace Davies
The fourth of the three graces
Too late for my last book
But still in time to help change
All of this ambivalent world
To grace and hope without end.

Preface

For some years past I have been dealing with the general and on-going issue of the largely troubled relationships of modern science and religion, principally the Christian religion. I have been doing this mainly in occasional university courses and seminars, and then at the odd conference and invited lecture. In the course of my researches in this area I had come to discover and to agree that the prevailing myth of the inevitable war between science and religion – as inevitable as war must be between blind faith, aka dark superstition, and rational discovery of the truth of things – was largely the invention of people like T. H. Huxley. It was an invention, and hence a false myth, engineered with malice aforethought towards the role of religion in 19th-century society, the pre-eminence of which they wished to replace by the pre-eminence of science.[1]

And I was of course aware of the equal and opposite antipathy to science on the part of fundamentalist Christians who insisted that what they knew, including what they knew about the creation of the world, was entirely dependent on unquestioning faith and owed nothing to a flawed and indeed corrupt human power of reason.

Then, in the course of one of those occasional fits of spring cleaning of my apparently self-cluttering workplace – these fits are very occasional indeed, separated by periods of up ten years, and brought on only when the clutter has reached such a degree that effective use of the workplace threatens to grind to a complete halt – I came upon a series of scripts of quite recent lectures and conference contributions that, as I looked more closely at their titles and contents, seemed as if they could form a coherent account of the relationships between modern science and religion. An account which seemed, moreover, to show increasing prospects of better and more promising relationships as time goes by.

Only two of these pieces have seen the light of day in other publications, and these are acknowledged in the relevant foot-

---

1. See James P. Mackey, 'Christianity, Critiques and Challenges' in Philip F. Esler, *Christianity for the Twenty-First Century*, Edinburgh: T&T Clark 1998, pp 25ff.

notes. One of these two pieces has later seen a fairly radical revision for purposes of presentation in a public lecture series, and
it is in this version that it appears below. The other is published
in the proceedings of the last of a series of conferences on
Religion and Postmodernism, the editors of which foresaw that
the papers would most probably be published elsewhere also –
perhaps because the editors and organisers of the conference
realised that Postmodernism is already fading into the obscurity
it so richly deserves? – and so they asked only that our contributions not be published before the publication of the conference
proceedings as a whole, which should then be duly acknowledged, as it is duly acknowledged below.

Only one new piece is added to this horde so recently excav-
ated from the chaos that precedes all creation. It has the quirky
title, *Concluding Unscientific Postscript*. A borrowed title, as those
who know Kierkegaard will realise. For he wrote on the relationship of the Christian faith to the Hegelian 'science' of his
time. A 'science' that, great as he was, I believe Kierkegaard radically misunderstood, largely on account of his own allegiance
to a fundamentalist position on the *sola fide*, 'by faith alone' principle of the Protestant Reformation of the 16th century. Here the
quirky title is used to introduce a piece in which an effort is
made to siphon off those conceptual confusions and illogical
procedures that continue on the scientists' side to hinder the
emerging, proper and more promising relationships between
science and religion, while acknowledging also the continuing
conceptual confusions and downright dogmatic falsehoods that
continue to hold back progress from the religious side of the debate. A new piece then, designed to help the dialogue forward in
order to redeem more fully the promise it already shows.

I have revised all of these pieces, mainly with the purpose of
excluding as much repetition as possible. I put the matter in
these terms because, by the nature of this case, a good deal of
repetition is bound to occur. Sub-themes of the topic of the
creation of universes that are central to the on-going debate between science and religion crop up in different contexts, usually
in connection with different forms and phases of that often acri-
monious debate itself. So, in the course of the revision I have
tried to make sure that where sub-themes are repeated, each
repetition of the relevant material takes it into engagements
with different positions or different moves made by those on

either side of the debate. So that there results for the reader – this is my hope – an accumulating return in comprehension of the material and in the clarity at least of the logic of the argument of the volume as a whole.

I am quite naturally not at all as confident about my present-ations of the science side of the conversation as I am about the philosophical-theological side. And that lack of dual-compet-ence counts as one of the main factors in what must seem to the outsider, and sometimes indeed to the protagonists themselves, as an argument between two groups of people, either one of which at regular intervals finds itself wondering if the other crowd are talking to them at all. In this connection I must ex-press here my gratitude to Michael O' Keeffe, Lecturer in Physics at the Waterford Institute of Technology, whose grasp of metaphysical issues I could see to be as sure and as clear as I had only to presume his grasp of the physics issues is. He guided me as best he could in understanding and expressing properly the scientific positions with which I had to deal. And this ac-knowledgment contains, of course, the usual codicil concerning those faults in that understanding and expression which must still be obvious, and obviously mine, in the work on which I now, with not a little relief, sign off.

The Festival of Lughnasa 2006

PART ONE

Is There a Distinctive Christian Doctrine of Creation?
Creationism and Evolution

CHAPTER ONE

*Evolution and the Idea of Creation-out-of-Nothing**

Take first the theme with the title, *creatio ex nihilo*, creation out of nothing. This is commonly presented as an account of the divine creation of the universe offered by Christians in the very early centuries of the church. The corresponding claim that creation out of nothing was at that time an exclusively Christian account of creation is then supported by surveys of the philosophical theology of 'the Greeks' in which the latter are said to describe the origin of our universe as some sort of emanation from or of the divine substance itself.

Now such a construal of this early natural theology of 'the Greeks', together with the exclusiveness of the Christian counter-claim, represents nothing short of a travesty of the true state of affairs. And its continuance to this day can be accounted for only by the presence of some particular need to keep on repeating a Christian prejudice. Despite the fact that this inevitably involves a repetition also of the ignorance that is both the necessary support and also the outcome of any prejudice. And also despite the fact that this Christian prejudice was not even shared by all of these early Christian theologians, many of the most innovative and illustrious of whom knew better.

In actual fact the phrase, creation out of nothing, may be legitimately applied to a wide variety of versions and forms of the accounts of divine creation found in the philosophy / natural theology of 'the Greeks,' and subsequently borrowed or duplicated by early Christian theologians. Differences between versions of this theme depended, for instance, upon the connotation one was able to give to the term 'nothing,' the no-thing, *to me on*. To give an example: prime matter, as distinct from matter already formed into things, was called *to me on*, the no-thing. So that the creation of the physical universe out of prime matter

---

* First delivered to a Religion and Science Symposium, on The Created Order, at the National University of Ireland, Cork, 13 March 2000.

could then be referred to as creation out of the no-thing. Yet this would not mean that prime matter existed before or independently of the creation of the physical and psycho-physical entities that make up our universe. For prime matter can never be found on its own: unless it is formed into a thing by some forming agency, it is not (yet) a thing; it is a no-thing; it comes into existence simultaneously with the formation of things. Just as darkness comes into existence with the shining of light (in the first Genesis creation story), as a limit factor on the edges of light. Just as, in Einstein's view of the universe, space-time (the new phrase for our physical universe) comes into existence as a structural quality of those newer fundamental forms in physics called 'fields'. In all of those cases it then seems from our worm's-eye perspective within the universe – and we then say it like this – that the forms known as fields are said to emerge in or out of space-time; whereas space-time emerges with the emergence of fields, then particles, and so on. The light is said to shine in or out of the darkness, whereas darkness emerges simultaneously with the shining of light; and formed things are said to be formed in or out of prime matter, whereas prime matter emerges with the formation of things: creation out of the no-thing, creation out of nothing.

Differences between forms of the many Greek accounts of divine creation out of nothing simply follow the differences between the imaginative forms used by myth, the abstract conceptual forms used by philosophy, and the mixed forms of philosophical commentary on stories of creation, particularly those stories found in the more mythic passages in Plato. As an example of these mixed forms, see Porphyry's commentary on the creation story in Plato's *Timaeus*. For Plato was now a *theios aner*, a divine man, according to his later academic followers, and his dialogues were then treated much as the Christian scriptures were treated by Christian theologians. Similar to early Christian theological commentaries on the story of creation in Genesis, in the course of which the mythic imagery of *tohu w' bohu*, of the void and the insubstantial, was commented and conceptually rendered by the abstract concept of nothingness. Justin Martyr acknowledged that Plato in the *Timaeus* gave the same account of divine creation as did Moses in Genesis. And even though he made this admission so that he could then claim that Plato stole the story from the earlier work of Moses, we are here nevertheless offered from the more informed and least prejudiced of early Christian theologians an example of the view that 'the Greeks' had much the same account of divine creation as they had. A view that in its abstract conceptual or philosophical form could be summarised as the creation-out-of-nothing view of divine creation.

And that is just one example of the fact that the creation out of nothing theme is seen to be well evidenced in the literature of Greek philosophical theology for those who care to take the trouble to study that literature. And not least from the time of Plato down to the time when Christians felt they needed a theological formulation of their faith and when, instead of creating a theology of their own, they adopted and only slightly adapted (change a few words and phrases, said Augustine of the Platonists, and they might as well be Christians) the thoroughly Platonised Greek philosophical theology of the Middle and Neo Platonic periods.

There were, of course, Greek theologies of creation that were characterised through and through by a dichotomous dualism of mind-substance and matter-substance – a form of extreme dualism, incidentally, that was neither invented nor subscribed to by Descartes[1] – in which matter, as a satanic or adversarial force, was thought to exist independently of God, and in that context to be co-eternal with God. What made these theologies false was not the eternity of matter, for that is quite compatible with creation out of nothing, but the independent origin and existence of matter, whether temporal or eternal, and its satanic status. Augustine, who was a devotee of such dualism in his early years, gives us an example of it when he quotes Faustus, his Manichee teacher, to say: 'There are two primary elements, God and Matter. To Matter we ascribe all maleficent, to God all beneficent potency.' And it is this theology, often attributed to a large and amorphous body called the Gnostics, that was rightly contested by the early Christians. But in this also the Christians were merely imitating 'the Greeks', for the orthodox Platonists also contested the dualist view of matter of the Gnostics, as the

---

1. See my *The Critique of Theological Reason*, Cambridge University Press 2000. Chapter One. Incidentally, extensive sections of this book thereafter take as examples of the main dialogue partners for contemporary Christian theology scientists who have made the more striking contemporary advances in physics and psychology.

very title of Plotinus's *Enneads* II, 9: 'Against the Gnostics' more than amply illustrates.

So an ignorant prejudice was perpetrated by some early Christians against all 'the Greeks', and it is right to ask that nothing be done to perpetuate it. It is bad enough if some theologians still perpetuate it. Too many theologians know little or no philosophy, and talk only to themselves, to priests and nuns and trainees for these professions, and to a small minority of laity. But educationists would surely not want to propagate this prejudice through any national education system. And this not merely because of a need to procure justice at last for those ancient Greeks. They are all long dead, and whether up on Olympus or down in Hades, they are presumably as indifferent now as they were then to what Christians, whom they presumed in any case to be an ignorant and poorly educated lot, were and still are saying about them. But the prejudice must be stopped in its tracks because of the damage this prejudice can still do to the attempt to maintain right relations between science and religion, an attempt in which our second level students now are invited to engage.

For 'the Greeks,' those foundational philosophers of the West – and here comes one of the extremely rare generalisations that can be at all countenanced in their case – the Greeks recognised and practised a complete continuity in their quest for knowledge (in Latin, *scientia*, science), from the level of that quest which current conventions segregate for the natural sciences, through the level now commonly called philosophy, and to the level then as now identified as theology. That last name, theology, was the name chosen by these Greeks for the level of the same inquiry into the nature of things that seems to observe and then to investigate, amongst all the agencies that reveal themselves in the natural universe, an ultimate and comprehensive creative agency that might well be the source of everything else and of every other process that is observed in action and investigated by human reason. This conventional name for such an ultimate creative agency was, and is *theos*, God, and the corresponding name for the rational investigation of such an agency, that leaves its traces in all of creation was, and is, the *logos* or rational account of *theos*, theology.

Now the travesty of the natural theology of the Greeks that Christian theologians still propogate has the effect of denying to the Greeks any true appropriation of the knowledge of God the creator from the creation itself, and it can correspondingly deny that the Greeks could be truly religious as a result of such natural knowledge.[2] They could be deemed at best to be some kind of idolatrous pantheists, depending on the advent of Christians preaching creation out of nothing to bring them to true religion. A second and consequent effect of the perpetuation of this travesty of the theology of the Greeks is the (deliberate?) spread of the impression that there are limits to what can be achieved by this natural quest for knowledge in its efforts to understand the nature of reality and of all the agencies and processes active on or in it, so that one could then act responsibly on or with these for the best outcome for all. And the spread of this impression is quite convenient for those who wish to draw a clear boundary between science and religion. So that proponents of science and of religion can hold friendly conversations, but only on the strict condition that neither tries to annex the other's questions or answers. The spread of this impression is especially convenient for those who wish to suggest that answers to questions concerning values, meaning of life and ultimate sources should be expected from religious folk, but never from the cultivators of natural *scientia*, of natural knowledge or science, as such.

Why was this prejudice against the Greeks produced by some early Christians? Probably because they desperately wanted to present themselves as people to whom God had revealed answers to life's deepest problems in this world, answers which in her infinite goodness and wisdom she had kept well hidden from everyone else. And maybe even, on the darker side, because they rather fancied the prestige and power that such favouritism would inevitably confer on them. Just imagine being the divinely chosen supreme authority on matters, not just of faith but of morals also. But then why, in God's name, would any Christian person, or indeed any truly religious person, want to repeat such a prejudice, with all of its attendant ignorance, to this day? Could it be that some still feel a need to find, amongst all the myriad questions posed by this always wonderful yet regularly appalling world, some set of questions, however small the set, so long as the questions are crucial, and turn out to be

---

2. See T. L. S. Sprigge, 'The God of the Philosophers: Faith and Reason', *Studies in World Christianity* 4.2 (1998) 149-172, for the case that natural theology can lead to a life of high morality and worship.

questions that others may well ask, but only religious folk can truly answer?

Or, to put the same question in another form: why would religious people want to create gaps in the world as naturally experienced by all of us? Gaps in the whole vast panorama of aspects of the empirical world and areas of its natural history, which on contemplation and investigation yield up what one hopes are ever more accurate and fulsome knowledge of all the agencies, powers and processes involved and of the promises for the future that these may be seen to hold out, on condition of course of a proper creative co-operation with these same agencies, powers and processes? And why would these spokespersons for religions then seek to fill these gaps with aspects or areas of the empirical created order that allegedly cannot yield such knowledge, theoretical and practical, through the natural means of human contemplation and investigation, but must await such knowledge being ferried to these areas from 'religion'? Why would people belonging to any religion do this, when they must then realise that the gods they thus represent are by definition gods of the gaps?

Take two examples from the religion and science debate as it is conducted today.

First, the very existence of the universe and of everything that has been, is or will be in it, in short, the fact that these exist at all, is an aspect treated separately from that aspect which concentrates attention on the processes by which all species of things develop into the kind of things they turn out to be. It is then said that evolution, hence science, accounts for the kinds of things that come to be, but only a religion can answer the question as to why any things exist at all, rather than nothing existing. And the answer to this question, which occurs in an alleged gap in the relevance or effectiveness of science, is: a once-off, long past, original act of creation-out-of-nothing took place, an act that by definition belongs to the agency of a divine being.

There is another version of this example of alleged gaps in the understanding and explanation of the universe that science cannot fill. This example focuses upon an entity characterised by awesome concentration of energy which 'then' exploded in a Big Bang. Science, it is then said, can account for the coming to be of all the kinds of things that come to be, from the point of the actual explosion onwards. But science cannot account for the bundle of infinitely concentrated energy itself. Now most scientists, like Stephen Hawking, do not like such space-time singularities; that is to say, entities or stages in creation that are singular in the sense that they do not appear to be governed by the same physical laws that account for all other entities and stages in the coming to be or our universe. For such scientists feel that they ought to be able to account for the formation of all things that come to be, including this infinitely dense entity that exploded. And they expect to do this by discovering the formulae or laws by which all, at all possible stages of the evolution of the universe, are formed, deformed, reformed and continuously transformed. It is precisely on this task that many physicists, like Hawking, are currently engaged.

Furthermore, there are some Christian theologians who talk about continuous divine creation, that is to say, a kind of divine creative act that continues long after the divine creation of the pre-Big-Bang entity. Yet it is not often clear quite how they conceive of this continuous creation. Does it mean that the very existence of all that comes into existence, as well as what each species of things comes to be during all of that subsequent evolution, is equally attributable to the same divine creative action? Or does it just mean that, after the Big Bang, the divinity simply maintains things in existence (whatever that means), and actively intervenes only occasionally, in order to repair, improve upon or add to what was originally created as one, unified universe? In the latter case, there still occurs a gap, the pre-Big-Bang entity or phase, which these theologians are happy to say scientists have not yet accounted for and explained, and hope they never will; and perhaps (please God) other gaps will always be found in the scientists' account of creation. Whereas in the former case the gap between creation and evolution would inevitably disappear; and with it the last of these 'gods of the gaps'. For the scientist does not see, and indeed it is difficult for anyone to see, a gap between the coming of any particular thing or of all things to be the things they are, and their coming to be things rather than no-things. And once that gap disappears, as it certainly seems it ought to, we are simply left with no option except to debate with the scientists the success or otherwise of their accounts of the coming to be of the entire fabric of reality that constitutes the universe. And in addition we are left, thankfully, at last, with the happy prospect of the death of that idiotic debate between creationism and evolution that currently wastes so much ink and paper from scientists and theologians alike.

A second example: it is constantly claimed that features of our universe such as values and meanings cannot be accounted for by scientists, and more, much more controversially, that these can then be supplied by a religion. Now a value consists in the propensity of a thing or process to enhance or to diminish (now as a disvalue) the well-being, the prospects of existence and life more abundant for all other things and processes that together with us humans make up the unified fabric of reality. Scientists can uncover all such values in the active relationships that bind together all the things and processes that make up the fabric of reality. Not least because of the fact that in a finite universe, all exist and live to some degree at the expense of others. Indeed the work of the scientist is itself a moral activity, concerned with value and disvalue through and through. For the scientist, like all the rest of us, learns all she will ever know by interacting with the universe of which we all are such intrinsic and indeed interdependent parts. So the scientist too must constantly consider what values or disvalues she is creating or promoting in the very practise of science itself. The very idea that the scientist just finds out how things work and what can then be done to or with them, and can then leave the use or abuse of her findings to the moralists, is a monstrous portrait of the scientist – a portrait immortalised for example in myths of the Frankenstein genre. Quite to the contrary, the scientist is by nature and vocation a moralist, whether atheist or not, and indeed the only fully competent moralist in the matter of how she is to conduct herself in pursuit of her own profession.

As for the meaning of life, that vague phrase seems to refer to certain prospects for human life which comprise such things as the enhancements that can make life worth living, the quality and perhaps quantity of life that is or can be on offer in this universe. The recent work of physicists such as Freeman Dyson, Frank Tippler, David Deutsch and others who try to extrapolate from present laws, energies and processes operative in the universe, in order to enable them to actually envisage eternity for life in the universe, and scenarios of varying value for such prospects and possibilities of eternal life, is quite enough to show what a shrinking gap is envisaged here for the insertion of a god. In short, the attempted portrayal of this gap filled with concerns for value and meaning, as a means of introducing a *deus ex machina* who will somehow supply both, is a doomed enterprise from the outset. And apart altogether from the fact that this enterprise is doomed, there is the well known objection from secular humanists to the effect that values and moral rules imposed by such a *deus ex machina* lose all of their moral character by the force of such imposition.

And why then, finally, in view of what has just been said, why do spokespersons for the religious side portray these attempts to introduce gods of the gaps – for that is what they are still doing, despite their own frequent denunciations of gods of the gaps – as a way of keeping religion (they should really say, theology) and science strictly separated by a strictly defined fence, over which theologians and scientists may, and indeed should hold friendly dialogue, even if it be only a dialogue about the line of the fence? Why? Because they operate such a narrow definition of the physical sciences that all questions of ultimate creative source of the universe, even though this creative source be portrayed as continuously creative within the universe, must like all questions concerning value and meaning of existence and life, be ruled out. But why insist on such an excluding definition? After all, the Greeks whose work is so dreadfully travestied in so much of this literature, saw traces of the presence of a creator-out-of-nothing, a most ultimate and comprehensive source of all that becomes and is, during their investigation of the nature, the *physis* – which gives us the term, physics – the dynamic nature and processes of the things that are.

These early Greek philosopher-theologians were and still are called physicists in their earliest emergence in Ionia. And they were later called *physiologoi*, the practitioners of the science of physics, with particular reference to the theological reaches of their science. For these reaches of that same science were then called theology simply because, as suggested already, *theos* was the conventional word for the most ultimate and comprehensive creative source revealed by its traces always present in a world always coming to be. And all of this was already well in place in Plato's time and after, arguably the high point of the development of the foundational philosophical tradition of the West. It was, all of it, also called philosophy, the love and pursuit of wisdom, because these physicists were in quest, not merely of some facts about agencies, powers and processes in and of the natural world. They were, if anything, even moreso in quest of the knowledge of how to live in and with that world, and in collusion with all of the agencies, powers and processes therein detected, so as to enhance and not diminish the existence and life of all. 'Science', after all, simply transliterates the Latin *scientia*, which means 'knowledge'. But since the knowledge in question concerns a strategy for living as much as an abstract truth – Heracleitus said it was a way as much as a theory – it deserves the title, wisdom. And those who give their lives to the pursuit of this wisdom are worthy of the title, philosophers, lovers of wisdom.

Scientists in the modern era also took themselves to be working in the area of natural philosophy; they were philosophers. Newton's masterpiece is entitled *Philosophiae Naturalis Principia Mathematica*. But did these modern natural philosophers, who advanced science so greatly, continue their observations and investigations into the agencies and processes ever active in the continuous creation/evolution of our empirical world, to the point of investigating values and meanings of existence and life? And did they proceed further to the point of seeking to observe and investigate an agency and process that could be considered the most ultimate source of all that comes to be, as it is operative within the universe? Did they extend their professional quest for knowledge into the area of moral philosophy, and into that further area of philosophy known as natural theology? Well, yes and no. The picture is a confusing one, probably because prevailing currents in the contiguous sea-scapes of science, philosophy and theology drove exploratory expeditions now in one direction, now in another, and were sometimes perceived to be driving expeditions in one direction when in actual fact they were really driving them in another.

On the one hand, the dominant brand of philosophy on our neighbouring island, and indeed in most of the English-speaking world for much of the last century, abjured metaphysics, the very area of philosophy to which its theological investigations traditionally belonged. Further, that particular brand of philosophy found itself facing serious difficulties when it tried to account adequately for moral valuation. Its rather silly fact-value dichotomy, and 'no ought from an is' mantra (any is from an ought, then?) exacerbated these difficulties rather then relieve them. Appearing under different titles at different times – Logical Positivism, Linguistic Analysis – this was really Humean scepticism masquerading as empiricism, and more recently exalting as the one access to truth (or, rather, to probability) what it erroneously described as the inductive methods of the natural sciences.

Further, in this and similar philosophical climates in modern times, what passed for philosophy of religion in most departments of philosophy, consisted mainly of the treatment of proofs of God's existence, of miracles, and of the problem of evil. Few if any of the philosophy teachers involved seemed to recognise the awesome logical oddity of the proofs of God's existence: one first defines the kind of being (a) god must be, and then sets out to prove, or disprove, the existence of this god. None seemed to pause to ask the obvious question: if we had never encountered this being that could be called god, how could we even begin to describe, much less define it? And if we had ever encountered it, what would be the point of this whole elaborate exercise of seeking to prove its existence? Philosophers may answer: we simply took over these alleged proofs from religious thinkers and then dealt them a dose of our own fine critical acumen. To which the only relevant response must be: no excuse; your own fine critical acumen should have been first to tell you that you should not take over for such lengthy and persistent treatment odd logical processes finessed and promoted by religious thinkers under undue influence of the secular rationalism of the so-called Age of Reason. Logically, and in essence, the use of god-language is a naming process for that agency that is found to be – if such is found – the most ultimate and comprehensive source of all that is and happens in our empirical universe; an agency judged on the same basis of evidence as any other agency and process that the scientist observes and investigates.

Miracles, next, are treated in philosophy of religion as instances of by-passing, if not defying natural agencies and processes and the scientific 'laws' which govern these. But to treat miracles like this is to treat them as instances of contributory proofs of a previously defined omnipotence; and then the same logical objection applies as noted above in the case of proofs in general. A member of that rare breed of historic Irish theologians, a scholar named Augustinus Hibernicus (and a man who did much less damage to Christian thought than did his North African namesake), once wrote a treatise in which he insisted that God acted through, not against, natural agencies and processes in the case of everything ever described as a miracle. Miracles, then, were very striking instances in which, say, natural processes were speeded up, but all still occurred as part of the great process of continuous creation, the 'law' of which could be simply described as bringing life out of death, and in the end bringing eternal life and existence for all.[3]

On the third area of philosophy of religion as commonly conceived and delivered, the problem of evil, two comments must suffice here. First, there is the oddity of treating the problem of evil as if it were a problem only for those who believe they encounter in this universe ever-present traces of a benign creator. But it is not at all a problem for philosophers of a secular humanist persuasion? Why not? Second, there is the ignorance so commonly found in both secularist and religious thinkers of biblical acknowledgment of the benign creator's complicity in evils of destruction and death. A complicity inevitable in view of the facts: (a) that finite things in a finite universe always exist at each other's expense. Stars die that living things might come to be; and in general, continuous creation, as the fire-symbol for the creator signifies, involves the continuous destruction of preceding forms so that the new and higher may evolve. And (b) that the divine creator always creates in and through the forms of the material universe 'first' originated, and in and through all those that evolved from these, with all that this implies for the co-creative responsibility of these created and evolved forms. Responsibility as to whether they act in an overall creative or an overall destructive fashion, and with all that this also implies for the correction of the image of raw omnipotence so often foisted on the creator. Implicitly quoting the exact terms and images of the creation stories that open Genesis, Isaiah (45:7) has God say: 'forming light and creating darkness, making peace/prosperity – shalom – and creating evil, I am the Lord who does all these things.'

The confusion painted into that whole picture of the modern treatment of the problem of evil then ranges from a portrait of an Amoral Omnipotent, like Bertrand Russell's 'relentless matter'

---

3. See John Carey, *A Single Ray of the Sun: Religious Speculations in Early Ireland*, Andover and Aberystwyth, 1999. Essay Two.

as ultimate source of all that evolves and is; that has no conceivable interest in values; that simply, quirkily but inevitably, eventually kills off all that lives and is valued in us; but yet deserves the name of god, as the supposed ultimate source of all. So Russell often capitalised Matter, in addition to calling it 'omnipotent'. And from there the confusion ranges so far as to affect the alternative image of the God of All-Value (or Love), who creates life in eternally higher forms, albeit inevitably 'burning off' in the creative fire the more matter-bound forms of all things in the course of the evolution of higher forms of the same universal life-principle. And who then responds to the human race, which is itself responsible for the highest percentage of the evil from which it itself suffers, simply by carrying on creating life and life ever more abundant for all. And the confusion then affects portraits of creators corresponding to all points in between these two extremes. Little wonder then that contemporary thought presents such a vista of expeditionary parties setting out in quest of moral and religious ways and destinations, only to be swept off course by dangerous currents running in every which direction.

And yet modern science provides us with copious examples of scientists as scientists, or at least in the name of science, or in view of the overall picture of reality painted by their science, going after values and even after views of cosmic creator-agencies. Scientists who specialise in the study of evolution, for example, frequently offer genetic accounts of the origin and function of valuation, and of the selection of, the sense of obligation to, and the subsequent realisation of certain values.[4] On the question of the meaning of life or of the meaning of the very existence of the universe itself, the very persistence of value is at least part of the answer to this question also, as is the work of the scientists mentioned already who work on the physics of eternal life. Of these, David Deutsch is undoubtedly amongst the most interesting, if only because he seems to address also, both as a physicist and a pioneer of the quantum computer, the question concerning the most ultimate and comprehensive agencies involved in the continuous coming-to-be of the universe. Briefly, in his version of a theory of everything, he names four processes or agencies that together weave what he calls, in the title of his

---

4. See Stephen R. L. Clark, 'The Goals of Goodness', *Studies in World Christianity* 4.2 (1998) 228-244.

book, *The Fabric of Reality*. These are: fields and particles and so on, as described in quantum physics; then the agencies and processes characteristic of evolution; then, thought; and finally, computation. The interesting question that remains after all of this is, of course, how you can have thought and computation, or even those evolutionary agencies and processes to which he attributes primarily the processing and development of inform-ation (knowledge again?), unless there is a thinker of the thought, a know-er of the knowledge, a comput-er for the com-putation? But one does not have to wrestle with the advanced physics of Omega point, as does Deutsch, in order to sample a modern scientific attempt to talk about values, meaning and ul-timate agencies involved in the coming-to-be of the universe. A classic and most readable essay by Bertrand Russell, 'A Free Man's Worship,' reprinted in a collection of his essays entitled *Mysticism and Logic*, provides a full account, in accordance with all that modern science tells us, of value and meaning, together with a depiction of the creative agent which throws up all the pullulating species and entities that make up this evolving uni-verse, as 'omnipotent matter rolling on its relentless way'.[5]

And it is only fair to add here that the erstwhile arid ground of British philosophy has recently become less arid, in that meta-physics is once again allowed to grow on it. Severe restrictions are placed on the kind and quantity of metaphysical thinking that may be allowed. Quine, for instance, insists that only those entities and processes necessarily referred to by the most ad-vanced scientific theories should be deemed to make up the fabric of reality. Or, as Strawson would put it, metaphysics must re-main descriptive, that is to say confined to articulating and making explicit what we see from within to be operating within the uni-verse, along the lines of the world's deepest and most universal existential structures. Metaphysics must never again become re-visionary, by pretending to compare and criticise different views of the world's existential structures as if from a viewpoint outside the world somehow made available to us. Well, that's al-right then. That reach of the scientific-philosophical investig-ation of the universe which is named theology or, even more

---

5. See my 'The Creator, the Scientist and the End of the World: New Options for the Conversation between Science and Theology', *Studies in World Christianity*, 6:2 (2000), 208-223; or see Chapter Three below, opening pages.

specifically, natural theology, always did and still does talk of traces of a divine creator acting on and through the very forms of matter that constitute the continual coming to be of that same universe. Although, in order to make that point clearer, we may well have to do a quick and efficient overhaul of our talk of transcendence, so that transcendence is never reduced to meaning that the creator resides outside the world. For that constitutes a spatial connotation of the term, transcendence, that really makes no sense. What we need to comprehend is the much more refined idea of an agent that continually transcends precisely by acting within and through all that is always transcended; a kind of mutual interplay of the images of transcendence and immanence that is, as a matter of fact, commonplace in the Bible.

So back to the question: why try to build a fence between religion and science? Is it because of an anxiety to prove religion to be a source of some necessary knowledge that cannot be accessible outside of your religion, or at any rate outside of a religion? And if not that, is it perhaps because of that false and misleading myth, created out of certain incidents, such as those surrounding Galileo, and Darwin and so on – incidents so distorted by the myth that they convey the impression that religion and science are natural enemies always involved in a war-to-the-death with each other? Unless they are kept strictly separate by allotting quite different features of the empirical world to each, together with the questions, or at least the answers that then belong respectively to each of these disciplines? That false myth of inevitable war-to-the-death between science and religion probably originated with Leucippus and Democritus, early Atomists and first-parents of the genre of particle physics that these ancient ones most probably intended should help to keep physics clear of the troublesome and often dangerous attention of religious protagonists. In any case, the myth of an inevitable war between science and religion was re-issued and most successfully popularised in much more recent times by Thomas Huxley and his supporters.

And it seems to still linger on in the minds of some of those who have otherwise, and quite rightly, understood that the disputes that arose around characters such as Galileo and Darwin, were not in fact disputes between religion and science at all. Rather were they disputes between, on the one hand, current establishment views representing that inextricable mixture of science and theology which has just been shown to be the norm in Western philosophy as a whole and, on the other hand, emerging views that, as part and parcel of the normal advance of knowledge, challenged some important part of the establishment view in each case. It is indeed difficult to avoid the suspicion that those whose strategy in the science-religion debate is to draw a clear boundary between the two still, however unwittingly, host and spread the viral conviction that if these two are allowed onto the same ground, they can do nothing but fight each other, and most probably to the death.

However that case may be, what must be said at this point of this seemingly intractable debate is this: science, philosophy and theology all proceed with the contemplation and investigation of this empirical universe of ours, the only one we know to exist. All three seek to understand the agents, powers and processes revealed in the course of that contemplation and investigation. And all seek, even more crucially, to understand how we humans, together with all other genera and species in this continually created order may exist in co-operation for the best outcome for all and the longest. If there are distinctions between the kinds of question these discipline-categories ask and answer, it can only be an artificial distinction for the sake, let us say, of specialisation – for the sake of the kind of efficiency that specialisation is commonly thought to promise. But specialisation – for specialisation has its well-known drawbacks also – must never prevent recognition of the fact that all three contribute to the same ultimately unified body of knowledge. So that significant moves in any one of these must ask significant questions of, and have significant effects upon the others. Theology, which specialises in the attempt to understand the most ultimate and comprehensive agent-source of all created and creative entities and processes, and the manner in which we may best co-operate with this, is immediately affected if science says, for example, that it does not see any revelation of such an agent-source operative in the universe, or that the most comprehensive agent-source it seems to see revealed there (Russell's omnipotent matter rolling on its relentless way, for instance) is scarcely worthy of the adjective, divine. For that kind of science-based position is a theological position – atheism, if that is what Russell's position must be called, is a theological position – and theology must deal with it. And theology may perhaps deal with such a position best by invoking other scientific cosmologies that are inherently critical of regarding something simply called 'matter' as the most comprehensive creative agency in the world. Just as a scientific cosmology might deal best with some theology that proves to be stubbornly dismissive of it, by invoking another theology together with the latter's competent criticism of the former.

Finally, then, as this piece began with a complaint about a travesty of the Greeks, it might as well end with an example of the dominant version of Greek theology of God the creator that was available to our earliest would-be theologians, to be used by them as a philosophical version of, and hence a commentary upon the story (that is, the myth) of creation in the first book of their Bible. This might also help to show how, if Christians did not lie in their teeth about this lock, stock and barrel borrowing from the Greeks, then as now, their relationship to the natural philosophers of all ages might have remained healthier, much healthier than the gap-makers and boundary-definers could ever make it, and that to the mutual benefit of all of those to whom God the creator is revealed in the book of creation as much as in the Holy Book. For the readers of both books are human and prone to error, and much less likely to err if they are each open to the advancing and often corrective insights of the other.

But first, and in order to understand how apposite our first theologians found the following piece of Greek theology of creation, it is necessary to reflect briefly on the terms and images of the Genesis creation myth. The root meaning of the Hebrew verb, *bara* (created), is to cut, as in cut out, shape, fashion, form. That term, *bara*, opens the creation story, and is repeated during the story, for the forming of the giant creatures of the sea and for all of the sea's teeming species, and for the forming of humans. The corroboration of this meaning is found then in the shaping of sky, land and sea, by the image of separating waters above from waters below, and then separating land from waters below. Then 'word' appears as agency in the creation through the repeated formula, 'and God said'. In all cases but one, this formula has God talking to herself, as if formulating a plan of action. 'Let the land produce vegetation; seed-bearing plants,' and later, 'Let the land produce living creatures according to their kind', and 'The land produced.' Nothing so far in this imagery to contradict or seriously qualify the image of forming. The land is formed so as to produce all flora and fauna, and the reference to seed-bearing suggests that these then form their future progeny.

Only one instance of the word-agency theme might be interpreted otherwise. 'Let there be light; and there was light': that does seem to be a word, not now in the mode of a plan for forming the various parts of creation, but rather in the mode of a command which without more ado brings about what it commands. Those who wish to imagine divine creation as an act of sheer, unqualified omnipotence fasten upon that latter mode of interpreting the word involved in divine creation as such. Those gap-makers who distinguish between the coming-into-existence of things from their coming-to-be-what-they-become, whether the distinction harp on the first blob of awesomely condensed energy, or on the origin of the mere existence rather than the kind of what comes into being, then reserve the forming mode of word agency to the (subsequent) evolution of all things. And then for both groups the specifically divine contribution to creation consists in sheer command, as an act of raw omnipotence. Little wonder, then, if divine guidance as to how creatures are to conduct themselves in God's creation comes to be interpreted entirely on the model of divine command requiring simple obedience. This, however, goes against the clear import of the majority imagery of the creation story: God creates by forming, and further by forming forms which will then form others: the land produces flora and fauna, and the seed imagery suggests that these then form their progeny, each after its kind. So God creates through these forms and thus raises them to the dignity of co-creators, with all the creative responsibility for the creation which that implies. This reaches the status of a moral responsibility for those species that advance to the stage of exercising their co-creative function with rational deliberation, as the Genesis myth goes on to suggest, by having God put our first parents in charge as stewards of creation.

That this is the correct reading of the Genesis myth is clear from the Platonised Stoic theology of creation that the early Christian theologians borrowed in order to present the creator God of Genesis to the Greek-educated world of late antiquity. According to this Middle-Platonist-Stoic theology, creation was by the Divine Word (*Logos*, as in the prologue to the Fourth Gospel). As well as Word, this divine creator was also called *pur technikon*, a craftsman-like fire, for all creation involves destroying old forms for new. And the creator of the cosmos was also called Zeus, of course, to whom the Stoic, Cleanthes, wrote a memorable hymn. Now in this philosophical context, Word was taken, not as in the mode of an expletive-like command, but as an intelligible formula. So the divine creator continuously creates a world ever changing and developing by forming all that is in it, not by simply commanding each thing or species to exist. Furthermore, this continuous divine forming proceeds with its creative activity by sowing in each species seed-words, *spermatikoi logoi*, sparks of the Word, to change the metaphor for a moment, if only to bring in once again the spirit-as-fire imagery of the act of creation. For it is through these seed-like forms fashioned by the divine Word in everything so created that the divine creator continuously creates everything that comes to be in all of this changing and evolving world. It would be foolish in the extreme to attribute to these Platonised Stoics a knowledge of the nature and processes of the evolution of our universe such as we have more recently achieved. Yet there can scarcely be a finer general formula for the manner in which the creative source fashions a universe that comes to be in ever more evolved forms over the whole of space-time, from the origin of the point of infinitely condensed energy to whatever end there may be, or not be (no space-time singularities need necessarily apply).

Dan Bradley, during the conference to which this paper was originally delivered, likened to sentences, or *logoi* as a Greek would say, in which words change in response to adaptation and so on, the genetic DNA that guides the reproduction and development of living things. A very fine metaphor indeed, not only for that which he wished to describe, but more generally for the continuing creative activity that engages a divine intelligence. So that the face of that divine intelligence that is turned to us consists in nothing less or more than the process of evolutionary coming-to-be of which science learns more and more with each passing day. With the more accurate image of creation as forming, instead of the image of creation as raw and irresistible command, the end of this increasingly confused and confusing debate between creationists and evolutionists must be in sight. Although, given the increasingly obvious stupidity of the species that names itself *homo sapiens*, as best illustrated by its permanent penchant for massacre of its own members, whether by warfare or by the deprivation of millions of the food that this earth can provide in abundance, the degree to which other forms of stupidity can be maintained should not be underestimated.

Christian theologians borrowed from these Platonised Stoics then, the 'natural law' means of moralising, the mainstay of all moralising in Catholic Christianity to this day. They could not have done otherwise. By continually creating through created dynamic forms, and never 'directly' creating apart from these, the power of the divine creator of all that is or ever will be enlists all of these forms of existence as co-creators, and is correspondingly vulnerable to their excesses and failures. This is particularly the case where a form of existence is capable of coming to know the *spermatikoi logoi* that form all other species of things; because the *spermatikos logos* that characterises such a knower of all others is a self-conscious intelligence capable of such levels of co-creativity as to claim it is made in the image of the divine creator herself. This form of life, exemplified in the human species and perhaps in some other species, then has moral responsibility for the exercise of the co-creative dignity conferred upon them, and this cannot be taken from them without destroying the very form of existence they represent. The divine creator is therefore most dependent on their willing and positive co-creative behaviour, and most vulnerable to their wilful and destructive activities – the suffering God, the compassionate God.

Yet as surely as Roman Catholic Christianity borrowed, with this theology of the Stoics, the natural law morality, just as surely did it gradually corrupt this system of moralising as the centuries passed, until the most obvious instance of such corruption occurred with the papal encyclical on contraception in 1968. In essence, the corruption consisted in superimposing the command image of divine creation on the forming image. So that it appeared that one could read off divine commands from natural processes; whereas in the divine forming through the created forms kind of imagery of creation, the dynamic created forms themselves had a co-creative part and responsibility in the evolution of natural processes.

Perhaps the best way to illustrate this point is by reference to Dolores Dooley's lecture to the same conference. Her theme, 'Playing God', examined uses of that complaint against others by theists and then by secularists. When theists complain that people are playing God, they have very much in mind God's omnipotence, omniscience and consequent commands. So that God's creating activities automatically set limits on what the rest of us creatures should be doing, or even planning. So, for example, God creates every human being, at will, and human beings, even if they do call their co-operation procreation, must do nothing to interfere with God's sovereign decisions in this affair. They must leave every act of intercourse open to God's creative designs upon the families involved. There is, of course, a 'safe period' in the woman's cycle, and that may be read, as the more irreverent (not Dolores Dooley) would put it, as God, whose sole decision it is to make new human beings whenever she wishes, allowing humans nevertheless to play a little Russian roulette with her, and then hope for the best, as they see it. But it is not their business to decide what is best in such matters, only to accept and do the sovereign will of the omniscient God as expressed in the course of nature. There could scarcely be a better illustration of the inferiority of the command model of divine creation and of the morality which follows from it. Or indeed, if only by contrast, a better illustration of the superiority of the forming model of divine creation in Genesis and in the Stoic moral system borrowed as a consequence in order to theologise the Judaeo-Christian creation myth. In the course of her lecture Dolores Dooley quoted Aquinas: 'We do not wrong God unless we wrong our own good.' Such a short sentence, yet such a clear endorsement of the true creative moral responsibility and dignity that the true creator God confers upon us. Our failures to respond to which she can only suffer, and in respect of which she can only carry on creating – carry on with the love-in-act of pouring out life and existence without limit or measure, making her sun to shine on the good and the bad, and refreshing with her rain the just and the unjust, as her uniquely beloved Son once so memorably phrased this most basic existential matter.

Intriguingly then Dolores Dooley described the manner in which secularist folk use the complaint about others playing God. They use it in the context of perplexity and anxiety before the uncertainties involved in our inevitable meddling with natural processes. In more picturesque terms, they use the complaint in the context of the fear that the Frankenstein myth might prove only too true. In such contexts, what is targeted is the tendency to move forward with genetic engineering, for instance, before knowing far more about the extremely complex processes involved. Moving, in short, through a premature presumption of near-omniscience, itself perhaps engendered by the truly awesome and still burgeoning successes of modern science. At what stage can we be sure that stem cells will not run riot as they do when they produce teratomas, and so on? But of course this is the kind of thing that theists with the correct theology of divine creation and the correspondingly correct assessment of the moral responsibility of co-creation should be saying. Not that a couple should accept infertility as God's will, rather than try IVF, but that scientists should know as much as possible about the extremely fine and complex ways in which the divine creator continually creates life, and life ever more abundant, before taking up the full responsibility of their very secondary but nevertheless real co-creative responsibilities. Not the first time, and not the last, for a secularist source to teach the over-confident religious how to address the most challenging moral issues of the day.

In summary, then, the theology of every religion and of any church has to engage directly and at all times with the science-philosophy-natural theology (even atheistic, indeed especially atheistic theology) in which the sum of formal human knowledge consists, and engage with it on the basis of shared questions and answers. Nor can the theology of any church or religion ever withdraw from this constant three-way dialogue merely by making claims to special revelations, however true such claims may prove to be. For special revelations cannot even take place unless God is already revealed in the creation itself. For if that were not the case, the proposition, 'I am your God, and I am now telling you this,' would make no sense. You would quite literally not know who or what was talking to you. And in any case, in the prologue to the Fourth Gospel, John clearly states that the Word through whom the world was created, and which John is just about to claim became incarnate in Jesus of Nazareth, always already enlightens 'everyone who comes into the world', or just 'everyone'; depending on how you translate and then punctuate the Greek sentence in John 1:9.

CHAPTER TWO

Natural Theology and Divine Revelation*

[*Abstract:* This paper is designed to offer the opportunity to think critically once more on the occasion of this Gifford Lectureship Centenary Conference, about what those concerned with the Gifford Lectures understand to comprise natural theology, as an aid to discovering both its real problems and its best prospects. In essence, then, and in memory of the understanding of natural theology by those who first designed the discipline, the suggestion is made that natural theology, far from being without appeal to revelation, is predicated upon the revelation of that which can be called divine in the history of the natural world. Indeed I would contend that an increasingly important part of contemporary Christian theology recognises that its revelation is as natural in medium as it is supernatural in source, just as the divine creative source of the universe is transcendent through its total immanence in the creation. And that this opens to Gifford lecturers the prospect of a dialogue with contemporary science that can carry it as far beyond the squabbling over the comprehensiveness of rival explanations, as that

---

*This paper was prepared for the Gifford Bequest International Conference at the University of Aberdeen, 26-29 May 2000. It was rejected; not surprisingly, in view of the fact that, if the argument is correct in its conclusions, then the Gifford Committee at Aberdeen might conceivably be suspect of misrepresenting Lord Gifford's intentions, and using his money to offer to the public a rather inferior product, over-attached to definitions provided by sometime suspect sources, and to proofs, disproofs, miracles and the problem of evil. This is not to suggest that Aberdeen, any more than the other Ancient Scottish Universities – St Andrews, Glasgow and Edinburgh – did often offer an inferior product in their annual Gifford Lectures; but it does point to a danger here, if the description of subject matter to be covered in the celebratory centenary conference were to be taken as adequate, not to mention prescriptive.

science has gone beyond proofs and probabilities towards cosmologies that try to do justice to continuing, if increasingly mysterious revelations from the depths of this mysterious universe, and from its furthest, if barely discernible prospects.]

It is my contention in this short paper that insufficient understandings of the very nature of the project known as natural theology exist, and that, due to a confluence of influences from both Christian theology and anglophone philosophy, these misunderstandings have become increasingly widespread. I believe that such insufficient understandings should be critically assessed, if only because for this occasion they can be shown to slow the proper advancement of Lord Gifford's great project.

Perhaps I could best approach this problem in the understanding of natural theology by juxtaposing some sentences from the Call for Papers for this conference, and some sentences of Lord Gifford's own, taken from the text of his Bequest. The Call for Papers: 'Natural theology is understood broadly as the search for knowledge of God without appeal to revelation. Topics include, but are not limited to, the nature of God, arguments for God's existence, and miracles.' Lord Gifford's Bequest: 'I wish the lecturers to treat their subject as a strictly natural science, the greatest of all possible sciences, indeed, in one sense, the only science, that of Infinite Being, without reference to or reliance upon any supposed special exceptional or so-called miraculous revelation ... they may freely discuss (and it may be well to do so) all questions about man's conceptions of God or the Infinite, their origin, nature, and truth, whether he can have any such conceptions.'

Now, beneath the surface similarities of these two sets of sentences, there are two disparities in particular upon which the problems in the understanding of natural theology may be said to centre. First, whereas the Call for Papers would deprive natural theology of 'the appeal to revelation', the Bequest would refuse it reliance only on 'special exceptional or so-called miraculous revelation'. The Bequest thus rules out appeal to the miraculous as revelation of the divine, whereas the Call for Papers explicitly includes miracles amongst the topics for natural theology. And, second, whereas the Call for Papers would have us discuss 'the nature of God', the Bequest would have us discuss only 'man's conceptions of God', or indeed, 'whether he can have any such conceptions'. Now I realise that that second discrepancy may seem at first blush to be minor, if not almost accidental. But I hope to show that it is serious, especially when taken in conjunction with the first discrepancy. How then do these discrepancies illustrate the difficulty in understanding the nature of natural theology?

Take the so-called arguments for God's existence. These have been regularly dismissed by the most eminent philosophers amongst Christian believers – by Kant, for instance, and Hegel – on the grounds that, as in the infamous Ontological Argument, some definition of divinity has first to be proposed to one who professes not to know of a god. But this puts even the proponent in the position of pretending that a definition of divinity can be formulated, that the nature of God can be described, before someone actually comes to know God. And the existence of a god so defined can then be argued into the agnostic's head, or into the head of the atheistic opponent, or indeed back into the proponent's head, if she later lapsed into agnosticism or even atheism. For why else would such a thing as an argument for God's existence ever be construed? I mean to say, if the proponent of such arguments knew God, however inchoately, from some disclosure of the divine within the common fabric of reality, why would she not simply attempt to lead the agnostic to the locus of such disclosure and simply say: what is disclosed here is what I'm calling God? Do you see it? And can we talk together as to how the source of these deep and dark intimations of a presence can be described without too much distortion?

One has to suspect that the reason why something called an argument has to be construed is that the disclosure, that is to say, the revelation of the divine within the fabric of reality, is discounted in general terms. (And divine revelation is dismissed in just such general terms in the Call for Papers, but not in the Bequest where only miraculous revelation is refused.) One has to suspect also that those philosophers are vindicated who criticise these descriptions of the nature of God, when used as prolegomena to the so-called proofs, as illustrated in the sequence of topics in the Call for Papers: first, the nature of God, then the arguments. Hegel called these propaedeutic descriptions of God's nature 'hollow, empty and poor'. Anthony Kenney much more recently showed the terms of such preliminary definitions, terms such as omnipotence and omniscience in particular, to be too confused and compromised for their service to any useful purpose in any conceivable argument for, or for that matter even against God's existence.

There is indeed all the difference in the world between discussing the nature of God without reference to revelation of the divine, and examining human conceptions of the divine in living encounter with a kind or level of disclosure which is there for all to see who would be willing to attend upon it. For such revelation, were it envisaged, could constantly improve upon or correct the descriptions that all give of what they see disclosed, and could let them discuss whether or not to call it divine. As things stand, such improvement and correction is scarcely to be expected, for what happens too often in fact is that the philosophers of religion adopt descriptions of the divine nature which some *soi-disant* recipients of divine revelation propose. These are often crude: God as omnipotent enforcer of his own will. And then arguments for God's existence are correspondingly compromised. Thus philosophers of religion, to whom natural theology naturally belongs, have the worst of both worlds: apparently refused access to divine revelation, yet in fact operating often with the most corrigible of definitions from its alleged recipients. There, in any case, is my reason for claiming that a choice of topic named 'the nature of God', in a view of natural theology which rules out revelation in the most general terms, causes difficulties for the understanding of natural theology. And such difficulties are not caused by a choice of topic named 'man's conceptions of God' (if any at all are available) in a view of natural theology which rules out, not revelation as such, but only miraculous revelation of the divine.

If I now concentrate upon this distinction, which occurs also in our juxtaposed texts, between reference to revelation unqualified, that is to say, general revelation of the divine, and 'special exceptional or so-called miraculous revelation' of the divine, it should be possible to come closer still to a view of the insufficiencies I here alleged in the understanding of natural theology.

I begin by observing that the discipline of theology was designed and named by the pre-Christian Greek philosophers. This discipline, its method and a very great deal of its content was then taken over by Christians from the time of the Apologists onward. The same Greeks, by Plato's time, had coined the phrase 'natural theology' in order to distinguish this from what they called 'mythical theology' and 'political theology.' Natural theology, *theologia physikon*, was the critical account of the revelation received in the course of the quest of the *physis ton onton*, the nature of things or, quite simply, the natural world. *Theos* then was the conventional word for the most comprehensive 'source'-form or forming entity that is active in the entire universe. Hence the *logos* of *theos*, theology, refers to the rational investigation and account of the source-former from the agency of which the whole of the formed natural world derived, and as that source was revealed through the natural world that derived from it: *theoplogia physikon*.

[*Theologia mythikon* referred to the same natural theology, as we also call it, but now apprehended and expressed in the imaginative, metaphorical, poetic language of myth. *Theologia politikon* referred to the official deliverances of those charged in society with the care and promotion of a people's religion – an inherently dangerous bed-fellow for natural theology because of the latter's innate penchant for philosophical critique, as the execution of Socrates in the name of the Athenian State Dogma concerning the gods of Athens more than amply illustrates. So for these Greeks, as for the Christians who borrowed their theology, revelation was a natural part of natural theology, and the latter could not be defined without the former.]

Nor did the Greeks confine this natural revelation of the divine to the physical entities and processes of the universe, to the exclusion of historical human beings. On the contrary, the divine source-fashioner of the universe, the *Logos*, as the Stoics named Zeus, a rational creator who continuously formed the world out of the no-thing, 'prime' matter, is immanently operative and hence revealed as surely in the minds of *logikoi*, that is to say rational animals, as in the continuous creativity we now call natural evolution. Humans are born with a 'spark' or seminal seed of the divine *logos* ever consciously active within them. In addition this creative Word at work is very specially and exceptionally revealed in and through particular human beings. And the revelation of the divine is more 'special' still in the case of those memorable individuals from history who have been particularly devoted to and adept at discerning the divine creator's ways with the world from their deep and constant meditation on such things, or from their greater openness to the light that comes to all from God.

These comprise seers like Diotima, the seer who enlightened Socrates, and purveyors of oracles, at Delphi for instance, as evidenced in stories about Socrates and in dramatic representations of his life and thought in Plato's *Dialogues*. Furthermore, Plato himself, in the course of the tradition to which he gave origin, came to be seen as a *theios aner*, a divine man, as Jesus also was in his time and place. It even came to be said of him, as was later said of Jesus, that he was born of a human mother, but not conceived of a human father. (That miracle was claimed because he was recognised as locus of a special revelation of the divine, and not vice versa.) And all of this was natural divine revelation, the necessary and sufficient source of natural theology. This Christians at times came as close to acknowledging as their own proselytising programmes could allow. 'They are great men and almost divine,' wrote Augustine of the Platonists, and he added for good measure, 'change a few words and propositions, and they might be Christians.'

Within this scheme of things, common to Greeks and early Christians, there could then be talk of special and exceptional revelations, to use two of Lord Gifford's adjectives. But these would refer to exceptional individuals within whose minds and lives the God ever active as continuous creator had elicited an especially clear or complete degree of receptivity of God's constant, immanent and revelatory creativity. And that revelation of the divine, that divine revelation, in both physical nature and human history, could be called simultaneously natural and supernatural. Natural in the sense already suggested, since human lives, thoughts, imaginings and expressions are an integral part of the natural world of entities and events, and increasingly crucial for the prospects of that world as time goes on. And simultaneously supernatural in that God, the immanent source of all of this revelatory creativity, simultaneously transcends all the entities and events of which the whole fabric of reality, the whole of the space-time continuum, consists.

So the early Christians helped themselves quite liberally to this Greek natural theology, and felt justified in doing so, since they allowed that the *Logos*, the Divine Word by whom all things were made, was made known to the Greeks before being made known, incarnate in Jesus of Nazareth, to them. And in doing this, they were not really contrasting natural with supernatural revelation and theology. They were simply claiming rather that the Divine Word who had not left the Greeks without witness to himself, was now revealed definitively in the very natural life, death and destiny of Jesus of Nazareth. Stories of miracles, in the sense of allegedly divine contraventions of the laws of nature, or the exceeding of natural powers, might be used by both Christians and Greeks to draw people to the loci of revelation of the divine in the history of the cosmos. But in and of themselves they did not constitute divine revelation. Taken in themselves, if instances can be verified, they belong in the context of proofs and arguments. (Indeed they are rather similar in logical structure to those controversies with scientists in which the religious side tries to maintain that scientific explanations of the cosmos and its evolution still leave some elements or aspects that only the existence of a god could explain.) And for all these reasons Lord Gifford was wise to rule out these forms of allegedly special and exceptional revelations that could be glossed as 'miraculous revelation', and to suggest that we rest satisfied with the human quest for the *physis ton onton*, the science of being, a science of being that might bring one to the borders of the infinite at any and every moment, from first to last. That was the scientific view of a reality that both in its immanent creative source and its transcendent goal dimly reveals the infinite, that Lord Gifford called the science of Infinite Being. (Lord Gifford might also have availed himself of the embargoes that the Bible reader must sooner or later come upon – embargoes placed by Jesus and then by Paul, on the anxious cries for coercive 'sign'-proofs of an essentially miraculous kind, that Jesus insisted were asked for only by an idolatrous generation. But one hesitates to quote such religious texts to pure philosophers.)

It is fascinating to find in recent Christian theology of revelation and in christology a recovery of, and a more thorough reworking of this original position on natural-yet-supernatural revelation and theology. I can but mention very briefly here the outcome of a sustained modern theology of revelation: the conclusion, namely, that divine revelation takes the form of history, and for Christians especially, the very human history of Jesus. Combine this with the insistence in the prologue to the Fourth Gospel on the fact that the Word incarnate in Jesus is the same creative Word that as such enlightens everyone. Enlightening everyone then, presumably, just by doing what this *Logos*-Word does: by continuously creating the natural world from which we all emerge. Combine the foregoing with a good Scottish exegete, James Dunne's well-reasoned interpretation of one of John the Evangelist's apparently contrasting and exclusivist statements. When John has Jesus say, 'I am the way, and the truth, and the life: no one comes to the Father but by me' (John 14:6), this according to Dunne expresses 'not an exclusiveness which denies God's presence anywhere else, but an inclusiveness which gives a means of recognising it everywhere'. And the point is even clearer now than it had been in the distant past: the point, namely, that the revelation of the creating and ever newly-creating God in the very human life, death and destiny of Jesus is, in the Christian claim, the definitive form, without flaw, of the revelation of the same God in the whole history of the universe.

How did this original view, so recently recovered in Christian theology, of the inter-relation of natural theology with revelation both general and special, come to be replaced by the very arguably deficient view which separates natural theology from divine revelation, if it does not also positively oppose these, one to the other? The short answer points to certain Christian views of sin. First, the early Apologists, not content with the positive case that the followers of Jesus benefited from the definitive reception in human history of the revelation of the ever immanent/transcendent creator and re-creator of all, strove to assert their superior knowledge of this revelation by additionally alleging demonic influences upon early Greek reception and representation of it. Second, and much more seriously, from Augustine onwards there developed a very controversial view of an 'original sin' committed by Adam and transmitted in the very course of conception to all subsequent human beings. And this sin came to be thought to blind all to the full, true and potentially salvific revelation of God, still potentially available for all in the continuously created and re-created world and its history.

The increasing influence in the modern era of a certain version of this Christian original sin theory, namely, the Augustinian-Calvinist version, created such an overwhelming impression of the effects of this sin in human nature that the very attempts which humans might make to come to a knowledge of this creator God, were themselves deemed sinful and could result only in idolatry with all its attendant evils. It followed then that natural theology and the 'natural' divine revelation which it presupposes, must be entirely replaced by a divine revelation which is supernatural; but now supernatural by contrast. That is to say, supernatural now in a ('miraculous') sense in which it is the contrary, not the co-relative of what can be simultaneously called natural in previous and other Greek and Christian theologies. Correspondingly, natural theology, if it is conducted at all despite this Calvinist embargo upon it, is judged from this viewpoint to be conducted in the entire absence of divine revelation. Now this is not the place or the time to construct a critique of an essentially Augustinian-Calvinist theology which would separate natural theology from divine revelation. But it is the time and place to say to the natural philosophers whom Lord Gifford had mostly in mind for his lectures, that they should have strong critical reservations about sharing such views concerning natural theology and divine revelation, even if they seem to share these views only on the single point of describing natural theology as the search for knowledge of God without appeal to revelation.

Finally, then, to a confluence of correspondingly questionable influence on this point from contemporary anglophone philosophy. When Logical Positivism gave way to Linguistic Analysis, the exclusivist claim of allegedly inductivist, hence reductivist science to provide the sole source of truth survived. And although there has since been much discussion of the readmission of metaphysics, the prevailing view is not promising for natural theology. This, of course, need not at all dismay philosophers in this tradition who also happened to be Christian believers, provided that they held to some version of the separation of divine revelation and natural reasoning, the extreme version of which is found in Calvinism to this day. A man like D. Z. Phillips might establish the meaningfulness of religious language along what he takes to be Wittgensteinian lines, yet decline to pronounce on the truth of any claims about God. Another might erect some form of argument for God's existence, but only as a disposable ladder to a divine revelation beyond the laws and powers of nature to exhibit and receive. Yet another might regard such argument to be hopeless on purely logical grounds – as Anthony Kenney does – yet allow that there might be vouchsafed religious faith experiences, again outside the powers of the history of our world to provide, and the 'miraculous' in some shape or form is centre-stage again. And most of these might be suspect of collusion, whether wittingly or unwittingly, with an essentially Augustinian-Calvinist Christian view which separates natural theology from divine revelation and, to one degree or another, even sets these in opposition.

But, once again, it is not necessary to subject these philosophers to this kind of interrogation and then, in line with modern Christian theology of revelation, to challenge critically some underlying assumptions. It is only necessary (but it is necessary) to point out that leading contemporary theory of science has left induction behind, as its principal, if not sole method of procedure. That is, if science ever was so bound to inductivist methods, as older British empiricists seem to have assumed. Instead intelligible formulae for processes of increasingly cosmological depth and comprehension are sought now by science, based upon what seems increasingly revealed to the (re)searcher from the still mysterious depths and the furthest prospects of the universe. So that some ensembles of intelligible formulae, if only in their extrapolative reaches, seem to paint again the kind of portrait of Infinite Being which persuaded the first physicist-philosophers of the West that they were, without more ado, theologians.

And so increasingly and on all sides Lord Gifford's idea of natural theology as the greatest of all sciences emerges. And with it the consequent and ever increasing prospect of the blessed relief of seeing the end of this hopeless hybrid, a natural theology with no disclosure to support it. And the end also of a natural theology that could face constant dismissal, or complete and utter up-staging from an allegedly 'supernatural' revelation, of the kind that Lord Gifford shrewdly perceived to be dependent upon miracle. Miracle in the sense of a happening entirely beyond the awesome range of the natural universe, and of our comprehension of it; if not in fact an event that ran contrary to both of these. A straw-in-the-wind pointing in the direction of the replacement of this hopeless hybrid with a more intelligible natural theology may be found in the works of contemporary Christian theologians, some even of a very Calvinist evangelical persuasion, who have begun to essay 'naturalist' theologies of their Christian faith.[1]

---

1. See, for example, Charley D. Hardwick, *Events of Grace: Naturalism, Existentialism and Theology,* Cambridge University Press 1996

PART TWO

*Advances in Relationships
between Science and Theology*

CHAPTER THREE

Science, Religion and
the Issue of a Godless Universe

No sooner had I offered this title to WIT for its public lecture series than I realised how dauntingly deep and worrisomely wide it was. No sooner, therefore, did I sit down to write the lecture than I discovered the urgent need to narrow the range of my topic and to find a way to make it, not I hope shallow, but at least accessible. So I decided to look over the last century, the 20th of the so-called Christian Era, to take for example two prominent scientists, one of whom made a significant contribution to our topic during the first decade of the 20th century, the other an equally significant contribution during the last decade of that same century, and to conduct in your presence tonight an experiment in assessing whether science and religion had diverged or converged in the last century of the last millennium. The two scientists are: Bertrand Russell and in particular a collection of essays of his published in 1910 under the title, *Philosophical Essays*; and David Deutsch and his book, *The Fabric of Reality*, published in 1997. Both of them, I should add, being confirmed atheists.

Bertrand Russell was indeed a great mathematician. His *Principia Mathematica*, written with the collaboration of A. N. Whitehead, has been compared favourably to Newton's great work of a similar title. He was also a great humanitarian, and a much published philosopher of the atheistic humanist persuasion. Those of my age or close to it will perhaps remember his leadership of CND, the campaign for nuclear disarmament, and his going to jail for that cause. He also fancied himself as a great writer, and in fact he was awarded the Nobel Prize for Literature in 1950. I'll read you some of his purple passages a little later,

---

* Delivered as part of a public lecture series at the Waterford Institute of Technology, 17 April 2003; a much revised version of an article, 'The Creator, the Scientist and the End of the World,' *Studies in World Christianity* 6:2 (2000), pp 208-223

and you can decide for yourselves whether that prize is always wisely bestowed. He also fancied himself as one of the world's great lovers; although some of the women whom he made the objects of his love (I chose that phrase carefully), are known to have demurred somewhat. But let us not quibble; two out of four ain't bad, and the two that ain't bad refer to strengths of his which are most relevant to our current interests.

In order to put Bertrand Russell in fuller historical context it is necessary only to recall once more the late Victorian Era and the designer war between science and religion instigated in particular by T. H. Huxley; a war, he would have us believe, as inevitable as war must be between reason and understanding on the one hand and, on the other, blind faith and dogma. This war, Huxley and his generals were convinced, had now come to its final and definitive battle, which for humanity's sake they had to win. And this they set about doing, even to the extent of borrowing for science the very nomenclature and symbolism once reserved for religion. They got people talking about 'the Church Scientific,' of 'the priesthood of science.' Huxley himself, in addition to being called 'Darwin's Bulldog,' was also known as 'the bishop'. The Natural History Museum in London was entitled 'Nature's Cathedral'. Huxley even succeeded in having Darwin buried in Westminster Cathedral, resting place of the great princes of the church in England, even though this was in direct contravention of the wishes of Darwin's family. He did not quite succeed in having a hymn to the creation replace all hymns and prayers to the creator at school assemblies, but he had a good shot at it.

There is not the time tonight to go more fully into this recent historical context for the myth of science dominant, so successfully promulgated by these Victorians, although one must always be conscious of the manner in which false myths, which manage to become popular, make people unconsciously prejudge matters without really giving any serious thought to them. Instead, straight to Bertrand Russell and to the divergence between science and religion, as clear as that between truth and superstition, and as wide as T. H. Huxley could ever have wished.

At the opening of one of Russell's *Philosophical Essays*, the essay entitled 'A Free Man's Worship,' Russell refers us to a scene from Marlowe's version of *Faust*. This is a scene in which

Dr Faustus is in his study, and Mephistophelis is describing to him God's motive for, and God's act of creation. And here is Mephistophelis' account of God's thoughts on creation, first of an angelic world and then of the world we know.

The endless praises of the choirs of angels had begun to grow wearisome; for, after all, did he not deserve their praise? Had he not given them endless joy? Would it not be more amusing to obtain endless praise? To be worshipped by beings whom he tortured? He smiled inwardly, and resolved that the great drama should be performed.

For countless ages the hot nebula whirled aimlessly through space. At length it began to take shape, the central mass threw off planets, the planets cooled, boiling seas and burning mountains heaved and tossed, from black masses of cloud sheets of rain deluged the barely solid crust. And now the first germ of life grew in the depths of the ocean, and developed rapidly in the fructifying warmth into vast forest trees, huge ferns springing from the damp mould, sea monsters breeding, fighting, devouring, and passing away. And from the monsters, as the play unfolded itself, Man was born, with the power of thought, the knowledge of good and evil, and the cruel thirst for worship. And Man saw that all is passing in this mad, monstrous world, that all is struggling to snatch, at any cost, a few brief moments of life before Death's inexorable decree. And Man said: 'There is a hidden purpose, could we but fathom it, and the purpose is good; for we must reverence something, and in the visible world there is nothing worthy of reverence.' And Man stood aside from the struggle, resolving that God intended harmony to come out of chaos by human efforts. And when he followed the instincts that God had transmitted to him from his ancestry of beasts of prey, he called it Sin, and asked God to forgive him. But he doubted whether he could be justly forgiven, until he invented a divine Plan by which God's wrath was to have been appeased. And seeing the present was bad, he made it yet worse, that thereby the future might be better. And he gave God thanks for the strength that enabled him to forego even the joys that were possible. And God smiled; and when he saw that Man had become perfect in renunciation and worship, he sent another sun through the sky, which crashed into Man's sun; and all returned again to nebula.

'Yes,' he murmured, 'it was a good play; I will have it performed again.'

'Such in outline,' Russell then comments, 'but even more purposeless, more void of meaning, is the world which Science (capital letter for science) presents for our belief.' What then does science propose that we believe about the world? Well, first and foremost, according to Russell, it proposes that instead of believing in the Creator God of Christianity or even in the Cruel Prankster God of Mephistophelis, we should believe instead that the source and origin of all this meaningless and purposeless world is what Russell calls omnipotent matter. Omnipotent – that is to say, almighty, all-powerful – in the sense that it brings into being, by the evolutionary processes which account for its continually creative activity, everything that was and is, that will be or ever could be. Omnipotent also in that these very same creative-evolutionary processes drive every thing and every species eventually to extinction. Omnipotent matter deals out life and death, existence for a time and extinction forever, with equal and utterly indifferent hands. And the resulting world is purposeless and meaningless, in that both the origins of new species and the extinction of old ones is accidental, due only to the happenstance of genetic mutation and of the collisions of astral bodies, of natural elements and forces. 'Man, too,' as Russell himself puts it, 'is the product of causes which had no provision of the end they were achieving. (Man's) origin, his growth, his hopes and fears, his loves and his beliefs, are but the outcome of accidental collocations of atoms.' For matter it was who, and I quote Russell once more, 'omnipotent but blind, in the revolutions of her secular hurryings through the abysses of space, has brought forth at last a child, subject still to her power, but gifted with sight, with knowledge of good and evil, with the capacity of judging all the works of his unthinking Mother.' (Capital M for Mother; omnipotent matter; Omnipotent Mother).

From there Russell goes on to explain how the long series of religions which have marred the history of human kind from the beginning, came about. Briefly, primitive humans, in terror before the indifferent cruelty of Omnipotent Matter, acting upon that oldest psychological principal – if you can't beat them, join 'em – worshipped naked might and reproduced it both in their rituals and in their social interactions. In ritual human sacrifice and in their penchant for war, sometimes bringing both together as when Israelites claimed that that old ethnic cleanser, Jahweh, in giving them a promised land (well, that was their story anyway, and for some of them it still is), ordered their army to kill off whole populations already in occupation of any land on their way to the promised land, and who dared to resist them. As Samuel, the prophet of the Lord, said to king Saul when ordering him to victory over the Amalekites: 'kill both men and women, the infant and the suckling child, the ox and sheep, the camel and the ass.'

However, as humanity evolved it found that it could envisage sets of values, ideals of goodness, truth and beauty by which human beings could deal, at least to each other, life and life more abundant, rather than deal always in death and destruction. But at first they made two connected mistakes about these values and ideals. They thought that these had come from the creator of all things, whom they were then beginning to see, not as some impersonal and indifferent power, but as an all-powerful person; and so they thought that these values could be realised to perfection, with the help of the almighty divinity and that he would protect themselves and their values into eternity; two connected mistakes. For as history progressed, these self-declared rational animals began to see that, over the whole range of the race, different versions of this allegedly absolute moral perfection and holiness before God encountered each other as rivals, and the persecutions and, yes, the killings in the name of rival religious cultures increased once more. It might have been as late in human history as Ludwig Feuerbach in the 19 century, the mentor of Marx in this matter, that human beings realised that these ideals of truth, goodness and beauty were all of them their very own creations and that, although the ideals would never be perfect in practice, they could still be sufficient to give meaning and purpose to human life which, like them, could never be more than temporary. Human beings being true to each other during their short span, and helping each other to goodness and beauty, however imperfect – that could make life well worth living.

And so Bertrand Russell concludes (and here is one of his purple passages on which you might well judge the justice of his Nobel award):

Brief and powerless is Man's life; on him and all his race the slow, sure doom falls pitiless and dark. Blind to good and evil, reckless of destruction, omnipotent matter rolls on its relentless way; for Man, condemned today to lose his dearest, tomorrow himself to pass through the gate of darkness, there remains only to cherish, ere yet the blow falls, the lofty thoughts that ennoble his little day; disdaining the coward terrors of the slave of Fate, to worship at the shrine that his own hands have built; undismayed by the empire of chance, to preserve a mind free from the wanton tyranny that rules his outward life; proudly defiant of the irresistible forces that tolerate, for a moment, his knowledge and his condemnation, to sustain alone, a weary but unyielding Atlas, the world that his own ideals have fashioned despite the trampling march of unconscious power.

I should end this part, as Russell did, with this fine passage. But I cannot resist the temptation to express it all once more in the form of a credo, a creed, for he did write of what Science (capital S) presents for our belief, and creeds have always served as the most succinct versions of systems of belief.

I believe in one Matter Almighty, creator of the starry heavens and the lean earth; of all that is, seen (humans and houses and hornets nests), and unseen (gravity and grace and goodness); and in her uniquely endowed child, the human race, continually crucified and dying for nothing, yet raising up values from the very graves of its kind, and ascending as far as may be to their fulfillment and its own; and I believe, not in a holy spirit, rather in an unholy but spirited body, and in its eventual death everlasting. Amen.

II

Now for my second scientist, David Deutsch. Deutsch made his reputation as one of the front-line exponents of that most sophisticated system to have emerged so far from the physical sciences, and to have continued to develop during the 20th century – quantum physics. But more importantly, he has proved to be one of the pioneers in the quest for the construction of the so-called quantum computer. This, in so far as I understand the matter is, or will be when they manage to construct one, a computer the raw material of which would be, not silicon, but the sub-atomic particles themselves, most if not all of which, together with their properties, are now thought to be known. For, awed as we may well be by the enormous calculating capacity and speed of the biggest and best of our current breed of computers, the material of these, silicon, has inbuilt limitations; crudely put, it is too gross a kind of material to allow virtually unlimited computational capacity. Whereas, if we could harness the particles themselves to serve as the base material for the computers of the future, then under certain circumstances that I must shortly attempt to describe, that capacity could become unlimited indeed. And a form of infinite and eternal possibility then enters the arena of human hope, and it enters our human range of visibility through the lenses of science itself.

The relevance of all of this for our topic tonight – the issue of religion and science in an allegedly godless universe – will all, I hope, become clear in a moment or two. But first I want to give one more reason for my choice of Deutsch over many other scientists who paint similarly advanced pictures of the universe – advanced, that is to say, with respect to Bertrand Russell's picture from the beginning of the 20th century. I have in mind here scientists like Frank Tipler, who in 1994 published a book entitled *The Physics of Immortality*, and Freeman Dyson who, some years earlier contributed to a professional and international scientific journal, *The Review of Modern Physics*, an article entitled, 'Life Without End'. (I shared a conference table with Dyson at an international conversation on science and religion in London, two years ago.) Now Tipler argues quite intentionally that a kind of contemporary scientific account of our universe that is quite similar to Deutsch's account, converges with and hence supports traditional religious beliefs concerning creation, although of course the cosmology of the latter would now look quite crude in comparison to the former. And Dyson was at the conference I mentioned precisely because he is well known as an advocate of the possible convergence of science and religion.

So I chose Deutsch as my dialogue partner tonight because, although all three, writing at the end of the 20th century, paint very similar pictures of the universe, at least in all the respects that refer to our issue, Tipler and Dyson might be suspect of tinting the picture painted in order to bring out some recognisably religious colouring. Whereas with Deutsch we find ourselves in the presence of a determined atheistic humanist of the Russell kind and, far from tinting his picture to achieve shades converging on the religious, he actually and explicitly argues that what might seem to us to be similarities between his picture of creation and religious ones, are in fact nothing of the sort. Deutsch certainly could never be suspected of even unconsciously bending evidence in the direction of convergence with and support for religious views of creation, and so we are on far safer ground with him in assessing the divergence or convergence of science and religion in the course of our chosen century.

First, then, in order to assess the scientific advances of that century, begin with a brief comparison of Russell's picture of the creation with that of Deutsch. For Russell, really there exists merely what he terms Omnipotent Matter, a simple name for all that is, if only because it names the single creative agent that, through 'accidental collocations of atoms', genetic mutations and other such accidents, large and small, brings into existence, and subsequently to extinction, all the individual entities and species that were or are or ever will be. Well, yes, there are things called thoughts, at least as long as *homo sapiens* exists. For, as Russell himself put it, Omnipotent Matter gives birth to *homo sapiens* in her blind hurryings through the abysses of space-time. And, yes, these thoughts are wondrous and varied: insights, creative visions, value judgments, the high emotions of love and the beauteous works of art, and they can make our lives well worth living. But to Russell and others who see the universe as he does, these thoughts are mere epiphenomena of a highly complex structure called a brain and nervous system.

Epiphenomena? Did I hear you ask? It is one of a group of words used by scientists, perhaps especially in medical science. These are words made up of two Greek words, difficult to spell and to pronounce, and thus giving the impression that the one who uses them must be in possession of deep and difficult knowledge about the matters in which he or she deals. Far, far too deep and difficult to be shared with us common folk in ordinary English. I suspect, partly because I did once learn Greek, that the opposite is often the case. If they really knew what they were talking about they could tell us in English. And it is therefore difficult to hush the rising suspicion that the real use of such esoteric terms is to hide for a little while longer the amount of ignorance that still surrounds their knowledge.

Take the one who says that thoughts are epiphenomena of brains. Epiphenomena is made up of the customary two Greek words: *epi* means above and around; *phenomena* means things that appear. So when you say that thoughts are nothing more than epiphenomena of brains you are simply saying that they are things that appear above and around brains and nervous systems; like sparks are tossed off and appear above and around electrically charged wires in contact. And this does not offer the slightest piece of information on what thought really is. If anything, it hints rather that the user of the term does not know what thought is, and does not really intend to make any further effort to find out. Perhaps because she wants to remain a pure materialist, untainted by things like minds, souls, spirits, intellects, or any such dross? Perish the thought! Why would any scientifically erudite person want to indulge in such verbal evasions, when it is perfectly obvious to any thinking mind – sorry, to any brain – that there is nothing but matter, that Omnipotent Matter is the source of all there is. And Russell's picture remains as single and as simple as I just said it was.

But the picture that Deutsch, in the name of the same science, paints of the same universe at the end of that same century, is a much more sophisticated and differentiated affair. I can give you a flavour of this much more sophisticated account of the make-up of creation, as we call it, in one single sentence of his: 'The fabric of reality does not consist only of reductionist ingredients such as space, time and subatomic particles, but also, for example, of life, thought and computation.' You can there see at a glance that Deutsch's universe is not as simple as a great conglomerate of matter – a space-time continuum made up of particles and forces, to which everything, and especially thought can be reduced so that the whole can be known by the analysis of matter and material processes alone.

Quite to the contrary, the creation, the universe is composed of four ingredients, as he called them, none of which can be reduced to any of the others, nor can any be reduced to being mere epiphenomena of any of the others. What Deutsch has called thought, what he elsewhere call knowledge, is just as real and at least as important an ingredient in the fabric of reality, as is the material space-time with its composite particles and forces. Correspondingly, there is for science not just one strand of investigation, analysis, explanation and understanding of reality. There are four such strands which Deutsch names as follows: first, quantum physics, for the investigation of the material sub-

strata of space-time with its particles and forces; second, epistemology, or theory of knowledge, for the analysis of knowledge
or thought; third, evolution theory, for the explanation of life;
and fourth, the theory of computation, and in particular the theory
and technology of quantum computers, for the understanding
and soon, perhaps, the human replication of what can only be
called the cosmic computing process. Put these four strands of
analysis of the four ingredients together, and it is possible to
construct the following crude portrait of cosmogenesis, that is to
say, of the continuous creation, also known as the evolution of
the cosmos, the making of the universe we know:

Knowledge, at first in the form of the laws of physics, is inscribed in the space-time continuum, in its constituent particles
and forces; in that all act in accordance with these laws, these intelligible formulae, in order to bring about, by Big Bang and expansion and cooling and star and planet formation, and so on,
the material universe we know, and to bring about life, or to
bring this universe to life. And with life to bring about a computed
increase, not merely in the speed with which knowledge is
transmitted (genetically that is), but an increase in knowledge itself, as computed creative adaptation of living species re-forms
reality for better or worse, adding beauty to beauty or ugliness
to ugliness, goodness to goodness or evil to evil, until a species
emerges that can co-operate consciously in all of this computed
creativity and increase its speed and efficacy to infinity, and into
eternity. And that species is – or at least if we behave ourselves,
it could be – us.

Take now a moment or two to look more closely at these final
prospects of that knowledge-driven or thought-driven and continuous computational creation of this universe that appears to
reveal the potential for eternal existence and life. That is to say,
take a moment to consult David Deutsch's version of what he
and scientists like him call their Omega Point Theory. For by
Omega Point they mean to refer to what religious folk call the
eschaton, the endtime, the endgame, the grand finale of this universe as we now know it. And it should then be possible to see
as well as we can see in one evening's work, the universe and its
creation as one advanced physicist at the end of the 20th century
sees it.

As he extrapolates from current scientific knowledge towards what can be predicted for Omega Point, Deutsch concen-

trates more and more upon *homo sapiens*, on our future evolution and increasing cosmic role. This universe, remember, is and has always been knowledge driven. It resembles, indeed it is, a computer, running programmes for ever-evolving adaptations to newly created conditions. A far cry from Russell's universe stumbling through atomic or genetic accidents with only the crude mechanism of survival of the fittest to guide it. Now our species can appropriate all the knowledge by which our universe is continually created, continually evolves. Furthermore, we can create little computers, and at our best we can create small instances of what is called virtual reality, that is to say, experiences of realities (flying a plane for instance) which are almost – that means, virtually – indistinguishable from the realities themselves. Now suppose that we had created our quantum computer of which Deutsch is one of the pioneers, and could reinscribe in new and creative forms on the very particles that currently carry it at atomic and genetic levels, the knowledge we have gleaned about the fabric of reality and the way in which it operates. What limits would there then be upon our ability to create virtual reality? You might expect Deutsch to reply to that question: none, no limits at all, for we would then be in charge of the cosmic computer itself and capable of programming it for indefinite evolution. But in fact he looks immediately to two potential and crippling limitations.

First, there is the limited state of our present level of intelligence and knowledge, and this is not just a limitation that can be overcome with time and industry. On the contrary, our intelligence is limited by increasing quantities of crass stupidity, of which it is only too easy to give examples. For one example, we are as a race irredeemably belligerent and have developed weapons of destruction with which in the course of one of our future wars we could literally destroy all life on this planet. But, to give an even more sinister example closer to home, we are presently engaged in causing increasing global warming and thus bringing about exactly the kind of conditions which, two hundred and fifty million years ago, destroyed 95 percent of all living species on this poor planet, starting with the most developed. *Homo sapiens*, Humanity the Wise, as we rather archly name ourselves, is quite possibly the stupidest species, and certainly the most destructive, ever to have infested this unfortunate universe.

But what if we did reverse our destructive stupidity? There is absolutely no sign of our being able or even willing to do this. Two of our Wise Leaders at this time, Blair and Bush, have decided that the best way to bring freedom to Iraq is to lay the country waste by use of their weapons of mass destruction, and to bring about, by their own ordinance, or by their criminal negligence to protect the people that they themselves deprived of government, police and army, the deaths of innocent men, women and children in a ratio of some six hundred to one, to the deaths of those who are actually fighting against them. Yes, yes, very well, but we're tired of talking of the Iraq war. It's happening somewhere over there, and we are here. Get on with the lecture. What if we did reverse our destructive stupidity some time reasonably soon? What if we did develop our intelligence and knowledge of the cosmic computer to the point where we could actually take charge of it, what then? And what, for this is more important than intelligence and knowledge, if we also developed the wisdom to use all of that knowledge for good rather than evil, and began at last to earn the title we confer on our species? Surely then there would be no limits to the quantity and quality of virtual reality we could create? Indeed, we could eventually drop the adjective 'virtual', for we would be inscribing our plans, our thoughts, our programmes on the actual particles of which the universe is made up. We should be creating an ever new heavens and an ever new earth, and with that we should be creating time, for time (I mean now the calendar time which depends upon movements of sun and earth) ... for such time, even the old theologians knew, is created with the world; the world is not created in time; and we could presumably create as much of time as we wished. True, but then at this point Deutsch enters the second limitation to be placed upon such enticing prospects.

There is not apparently in this universe to which we humans have access, enough matter and energy to make possible the unlimited memory capacity and the unlimited number of computations that this computer, now run by our vastly more intelligent and responsible successors, would need in order to allow the eternal creation of our new heavens and new earth. Except in one unique set of circumstances. These are the circumstances of a space-time singularity (that means a kind of event that happens only once in space-time, and is governed by a special set of physical formulae); and the space-time singularity in question here is the Big Crunch. Those scientists who believe that the universe originated in a Big Bang, an explosion, tend to think that it will end in an equally violent Big Crunch, an implosion.

You have had a story of the creation of a universe at the beginning from Mephistophelis. Now here is a story of the creation of a universe at the end of the world from David Deutsch. (In this awesome, violent implosion of the end-time, at Omega Point) 'the shape of the universe would change from a 3-sphere to the three-dimensional analogue of the surface of an ellipsoid. The degree of deformation would increase, and then decrease, and then increase again more rapidly with respect to a different axis. Both the amplitude and the frequency of these oscillations would increase without limit as the final singularity was approached, so that a literally infinite number of oscillations would occur even though the end would come within a finite time. Matter as we know it would not survive: all matter, even the atoms themselves, would be wrenched apart by the gravitational shearing forces generated by the deformed spacetime.'

Now before you think of asking me any questions about it, let me state quite categorically that I would not recognise a three-dimensional analogue of the surface of an ellipsoid, if I tripped over one in the corridor: I gave up maths-physics my second year at university. But I do understand enough of the picture he paints of the infinite and violent oscillations of the death-throes of the present universe, to take it that the limitless energy developed in the circumstances of that space-time singularity would be more than sufficient for the infinite memory capacity and the infinite number of computations required for the creation, out of the shards of this one, of a new universe. As the second half of Deutsch's story of the violent, dying seconds of the present universe puts it:

These peoples' minds will be running as computer programmes in computers whose physical speed is increasing without limit. Their thoughts will, like ours, be virtual reality renderings performed by these computers. It is true that at the end of that final second the whole sophisticated mechanism will be destroyed. But we know that the subjective duration of a virtual reality experience is determined not by the elapsed time, but by the computations that are performed in that time. In an infinite number of computational steps there is time for an infinite number of thoughts – plenty of time for thinkers to place themselves into any virtual reality environment they like, and to experience it for however long they like. If they tire of it, they can switch to any other environment, or to any number of environments they care to design. They will be in no hurry, for subjectively they will live forever. With one second, or one microsecond, to go, they will still have 'all the time in the world' to do more, experience more, create more – infinitely more – than anyone in the multiverse will ever have done before them.

Now there are a great many facets of this fascinating story of the end-time of this universe that would merit much discussion; but the single facet of the story that concerns us here is that which describes intelligent beings, and not just intelligent but omnipotent, omniscient and omnipresent beings, as creators of whole and eternal universes must be, 'creating more,' as Deutsch put it, 'infinitely more than anyone will ever have done before them'. And Deutsch knows as well as we do that beings of such high intelligence that they can create universes, do commonly qualify as divinities. He is correspondingly anxious to deprive us of any support for belief in creator divinities, by bringing home to us the differences between his end-time creators and our Christian Creator God in particular. He insists, for instance, that the Christian creator is one god, whereas his end-time creators are many, a veritable company of possible successors of ours; though vastly more intelligent and wise than we now are. And that is a fair point. Although, as the briefest re-telling of the Christian creation story at the beginning of the Gospel of John – 'In the beginning was the Word and the Word was with God, and the Word was God (Word, *Logos* in Greek, meaning an intelligence). He was in the beginning with God; all things were made through him, and without him was not made anything that was made' – as this re telling of the Christian story would seem to suggest, God creates through this Word, and there is also the Holy Spirit; so that unity of God and multiplicity of persons does not form the kind of completely contrasting picture that Deutsch thinks it must.

Deutsch does not make much if anything of the point that his end-time creators of a universe actually evolve out of the present universe and, furthermore, that they create their eternal universe out of the particles of this one, perhaps only a mini-second before these particles also are torn apart. He makes nothing of the point that these do not in any case create a universe out of nothing, and are not themselves uncreated. But, then, when you come to think of it, maybe he was wise not to make too much of that point. For that might only prompt us to ask him to say a little more about the creation of this present universe from which these end-time creators evolve and out of the material of which they create eternal universes. We might point out to him, for example, that it is infinitely knowledgeable, infinitely thoughtful persons who create the eternal universe out of the particles of this one (thus making this universe eternal?), whereas the impression given throughout his book is that it is something called knowledge or thought that continually creates this present universe.

For of the four ingredients that Deutsch describes as, quite literally, going into the making of this present universe, knowledge or thought is without qualification the one which contributes the truly creative action in continually making with or in matter all the things and species that make up this universe. Call to mind again the four ingredients: knowledge or thought, life evolving, computation and quantum particles. Notice that three of these have to do mainly with knowledge or thought. First, knowledge itself of course; second, life as evolution, which he thinks of mainly as a process for transmitting and increasing knowledge; third, computation which again is both a process for increasing and a process for applying knowledge; and indeed, when you come to the fourth one, quantum particles are not devoid of knowledge, for they have their properties which direct their actions and interactions according to the laws of physics somehow inscribed in them. So you could say, in rough summary, that world in the making has but two ingredients: one called knowledge or thought and the other called matter, which we presently imagine to be made up of tinier and tinier bits and pieces, but which some contemporary scientists, as well as philosophers as ancient as Aristotle, suspect must reveal a deeper level that is not in particle form at all. However that may be, knowledge or thought would certainly seem to be the pro active, creative ingredient in all the making of the world from beginning to end, from time to eternity, and matter the infinitely plastic, passive or receptive element, which might well be simply co-created in the course of the creative formation and continuous re-formation of all things and species that make up the universe.

But now, and here is the catch for Deutsch, the new universe created at the end-time of this one requires knowledgeable and thoughtful agents to create it; yet the impression is given that knowledge or thought does all the creating of the present one as if knowledge or thought were agents. Except that knowledge and thought are nouns for activities of knowing and thinking, and for the results of these activities. They are not nouns that name agents. Furthermore, unlike the grin on the face of the Cheshire cat, which in a well-known children's story written for adults, hangs about in the absence of the Cheshire cat, knowledge and thought do not hang about in the absence of knower and thinker. And they certainly, during such absences, do not create anything. So the upshot of the comparison of Deutsch's formula for the creation of the new universe at the end-time, with his hinted rather than explicitly stated formula for the continuous creation of the current universe, is surely this: that we seem driven to conclude that if the continuation of or from this universe to be created at its end time has knowledgeable, thoughtful persons as its creators, though these evolve from this universe and depend on using its already formulated particles, then the creation of this present universe from which such extraordinary end-time people and projects can come, must equally involve an infinitely knowledgeable agent-creator, the presence of whom no amount of reference merely to abstract terms like knowledge or thought can for any length of thoughtful time hide from our perspective and questioning minds.

There is a great deal more to be said concerning the clear convergence between, on the one hand, Deutsch's knowledge-driven creation of this universe, together with any new universes which may yet be created out of its end-time destruction and, on the other hand, the knower-driven creation, both of this present universe and of a new heavens and a new earth, that we hear about in Jewish, Muslim and Christian creation stories and their attendant theologies. There is, for instance, in addition to the Christian conviction that such cosmic creation was initiated by the knowledge-bearer known as the Word of God, the belief that this divine creative activity is continuous in and through a process that we from our end perceive as evolution. There is, furthermore, the belief that as evolution seems to show, God through the Word continues to create the world from within, continuously creating in and through the co-operative agency of all creatures. And most specifically through the agency of that species believed to be made in the image of the creator God, *homo sapiens* (God help us all). So that Paul in the Christian scriptures can picture the whole of creation groaning under humanity's option for destruction over co-creation, and now awaiting its share in the liberation of these human sons of God, when they opt once again for co-creation, in order at whatever cost to themselves to make up for the destruction they continue to cause. So that they can then take their full part in the final restoration of creation to the status of an eternity of existence in a new heavens and a new earth; much as Deutsch, unwittingly no doubt, describes them as doing at the end-time.

But to argue convergence in such detail would require at least another lecture devoted, this time, to the Christian stories of creation and to their detailed interpretation in modern theology. And that could not be done in a talk devoted to assessing convergence from the science side. But I think that the analysis offered so far does establish at least this: that in the course of the century just past, science and religion, in the matter of this most basic issue of cosmogenesis, did actually begin to converge once more, after undergoing a very distressing, if somewhat forced divergence at the hands of T. H. Huxley and his followers. Of course, as your poor hurting heads can probably by now testify, it takes a difficult piece of analysis and argument to become convinced of that convergence. Indeed, I must confess that having done my best to engage in that analysis and argument, with as much intelligence, truth and honesty as I could muster, I myself was still inclined to wonder if my conclusion concerning a new convergence was correct.

And then, on the very morning on which I finished the first draft of this talk, I opened my *Irish Times* and saw that Richard Dawkins was coming to Dublin to give the Science Today lecture at the RDS. This is Richard (*The Selfish Gene*) Dawkins, a research biologist who holds a chair in Oxford as Professor of the Public Understanding of Science, but who seems to spend most of his time persuading the same public of old Huxley's thesis: namely, as the advertisement for Dawkins' Dublin lecture put it, that science and religion are the great irreconcileables. So when I saw from the accompanying article in the *Irish Times* that the substance of the lecture he had left Oxford to bring on tour, was an all-out attack by this militant, indeed missionary atheist, upon scientists who were now supporting from their side a new convergence with religious belief, and when I saw from the article how many more of these there were than I had known about or realised, I decided that there must be a good deal more going for the case I have just argued for you, and that I can entrust it with confidence to your own thoughtful consideration. If Dawkins is that worried about it, then the analysis I have offered to you, and the case I have argued, has a good deal going for it indeed.

CHAPTER FOUR

Postmodernism, Transcendent Immanence and Evolutionary Creation*

The aim of this piece is to bring together a critical account of the Christian doctrine of divine creation with an equally critical account of some relatable elements of the best of postmodernist thought, and then to see, to put the matter perhaps too bluntly, if prospects of collaboration emerge or if we are still in the more familiar war-torn territory of recent decades. Now this brief account of the intended content of this paper does not contain any mention of the term 'transcendence', even though that term defines the topic of this conference and occurs also in the title of this paper. This is because, as David Wood's contribution to this conference amply illustrates, all known deployment of that term and its derivatives could be adequately accounted for in a godless universe. And, further, because many all too common usages of the terms 'transcendence' and 'the transcendent' as straightforward ciphers for divinity are examples of the sin of onto-philosophy, as that latter term is to be defined and argued against later. Transcendence then will be talked about, as the theme of this conference requires, but only in such a critical manner, and in the context of discussing from the point of view of a particular, indeed foundational piece of Christian theology, namely, the Christian theology of creation, the relationship of religion and postmodernism, as required by the general theme of the series of which this conference is the fourth. In sum, the task undertaken by this piece would be all the less compromised if the term 'transcendence' were not used at all. So then to the Christian theology of creation.

---

* Delivered to the Fourth International Conference on Religion and Postmodernism at Villanova University, PA, 18-20 September 2003. Published in John D. Caputo and Michael J. Scanlon (eds), *Transcendence and Beyond*, Indiana University Press 2006, and included here in a somewhat revised version by kind permission of the editors.

I. CREATION

The Christian doctrine of creation, as commonly presented, is
frequently bedevilled by two related and equally questionable
assumptions, namely, and first, that this act of divine creation
consists, or rather consisted in a one-off act, or a one-off series of
acts of naked power, imagined as acts of unconditional and irresistible command. From this peremptory command there resulted
what a contemporary scientific cosmologist would call a space-
time singularity. And by a space-time singularity is meant a
state or stage of the universe brought about by processes different from those that govern its continuous coming-into-being as
experienced in all other stages. The space-time singularity that
resulted in the Creator's command is then thought to have consisted in some inconceivably dense globule of matter-energy,
and it was from this, on the occasion of an explosion called the
Big Bang, that in turn the universe we now know derived.[1] The
second assumption is this: that this doctrine of divine creation
can be caught properly and accurately only in a highly abstract
conceptual formulation of the theme: *creatio ex nihilo*. The standard unpacking of the connotation of that abstract conceptual
formulation of the phrase yielded little, if anything at all, in addition to the following two negatives: that the world was not
created out of parts of the divine being or substance itself, that is
to say, out of the divine *ousia*. And that it was not created out of
any material that could be thought to have co-existed with, or
even pre-existed the divine act of creation. Despite the almost
entirely negative yield of this abstract conceptual formulation
those mythic formulations of belief in the divine creation of the
universe so freely available in the age of Genesis were commonly
refused consideration, on the grounds that they represented
nothing more than illusory and unedifying imaginings from the
wayward childhood of the race. The creation stories in the open-

1. On the idea of a space-time singularity and scientists' natural antipathy towards it, see Stephen Hawking, *Black Holes and Baby Universes*
(London: Bantam Books 1994, esp pp 39-41); on the alignment of space-
time singularity with the once-off command model of divine creation
and the alignment of the forming model with the conviction of continuous creation, and hence the possibility of natural theology as a possibility for scientists also, as physics opens quite naturally onto metaphysics
for them; for all of this there is not here the space-time necessary. But
see for example my interrogation of David Deutsch's *The Fabric of
Reality* in my *The Critique of Theological Reason*, pp 120ff.

ing chapters of Genesis were then loudly proclaimed not to be myth, and their obvious similarities to the creation myths of surrounding civilisations in the Ancient Near East were explained (away) as the clever inclusion of elements, symbols and themes from this ambient mythology precisely in order to counteract the false impressions and bad examples inevitably conveyed in their own context, by including these elements now in the context of the serene omnipotence of a divine creator who brought all things into being, each in its own place and in all its pristine goodness, without struggle or compromise.

If that second assumption above, concerning the adequacy of the abstract conceptual formula for divine creation and the inadequacy of mythic formulations is considered first, then it does not take a great deal of critical and literary critical acumen to see how eminently challengeable it is. The case against the assumption can be made in both general and particular terms. In general terms, it can be maintained that imagination is the prime heuristic 'faculty' of the human mind; and everything from the image and the metaphor, through the poem, to the fully-fledged myth forms both its tools and its express products. In particular terms, the Genesis creation stories can be read wholly and precisely as myth, so as to reveal a fuller and more proper understanding of divine creation than the abstruse conceptual formulation of the theme, *creatio ex nihilo*, could ever attain.

There is not the time on this occasion to argue in general terms for the permanent primacy in the human quest for knowledge of concrete imagery over abstract conceptualising. So, in order to take the shortest cut to that desired conclusion, focus for a moment on an epistemological principle accepted by Aquinas, though by no means first invented by him: *nihil est in intellectu quod non prius fuerit in sensu;* nothing is in the intellect that will not have been present previously in sense perception. Such a strong and unqualified claim for the place of imagery as the prime heuristic power of the human mind has seldom exercised its due influence in Western epistemology. Indeed it reached its nadir in the course of the Enlightenment, the arid rationalism of which has left few if any academic disciplines undamaged to the present day. Nevertheless, it is true, as the principle in its most unqualified sense implies, that the image, and thereby the imagination, in the attempt to construe reality at all its ranges and levels, in response to reality's most concrete and comprehensive self-revelations, is ever and always the prime heuristic process. For imagination deals in living, organic or at least structured wholes. Whereas the analytic function of the mind, which abstracts certain features of such whole structures and organisms, thereby forming ideas or definitions, and then perhaps tries by synthesis to put these back together again – usually with as much success as can be expected by the one who tries to re-assemble a frog after vivisection – deals in aspects, features, parts, qualities and quantities, which are never in fact found wandering about on their own.

And if the universe is itself a unified structure, perhaps in some respects reaching the status of an organism, in which all things act and re-act for the maintenance and, in addition, the further development of each and all, and if the would-be knower in fact learns all she will ever know precisely through that co-operation with other entities, as the best of pragmatist epistemologies in essence maintain, then the profusion of imagery required for the most basic cognition of such a universe of being will of necessity be very complex indeed. For instance, the knowing-by-engaging-with other persons and things will involve envisaging how certain developments can work to the benefit of one and all. And that simple fact engages automatically the creative and indeed the ethical exercises of imagination. And the optimal form of the complex imagery, which then forms the seed-bed of all possible knowledge, may well then be the story. And the conventional term for stories that operate at the broadest communal or cosmic, that is, metaphysical dimension is: myth.

In addition, any particular heuristic excursus of the human mind is enfleshed in a human body; and it is inevitably embodied also in the language in which its research is essayed and its results expressed. So that there is commonly revealed in all human discourse an inextricable mixture of image and abstract idea or – perhaps a better way of putting this point – a kind of continuum in which imagery and abstract ideation fade almost imperceptibly into each other. For this distinction between image and abstract concept is itself an example of abstract analysis. And it is as liable as any other analysis by abstraction to fall foul of the characteristic temptation of its kind. The temptation, that is to say, to take to be separate and independent entities what it abstracts from reality in the course of the mind's analytic efforts. Whereas what actually and always exists in all human discourse is the more or less imaginative and the more or less conceptualised.

Yielding to this characteristic temptation of the analytic mind produces at one level of our analysis of reality what is often nowadays called onto-theology. Presumably it is commonly called onto-theology because those who identify instances in which the abstractions of the analytic mind are reified, treated as separate and independent entities (e.g. *Logos*, Reason), commonly assume that such reified concepts fall within the traditional category of the divine, or within the traditional category of the metaphysical which these days is often presumed to be, if not coterminous with, then frequently conflated with the category of the divine.

But there are many examples in modern philosophy of what we may have to insist instead in calling onto-philosophy. For these involve the same reifying of what are in reality analytic abstractions; that is to say, they involve the making into things-in-themselves what are elements or facets of things that are abstracted for the sake of analysis, but are never in fact found wandering around on their own – like the concept of a line, a length without breadth or depth, in physics or geometry, for example. Yet such reifications of analytic abstractions do not result in anything we would normally call divine. And it therefore makes for a neater, less biased account of these matters if we place the instances of what are so often called onto-theology into the more comprehensive category of onto-philosophy, if only because we could then reach the more equitable and defensible position of being able to tar with the same critical brush both the psychologists who prattle on and on about 'the unconscious', for instance, and the theologians who prattle on and on about 'The Intellectual Principle', 'The Absolute', 'The Transcendent', just so you will be sure to know that you are talking to real theologians here.

The non-theological instances of onto-philosophy do go largely unnoticed, and certainly uncriticised in the Age of Heroic Materialism, as it has been aptly called. Take the example of the analysis of the human being into the ideas of spirit and body, mind and matter, inner and extended substance. The abstracted terms of this dualistic analysis are regularly treated as if they named two separate and distinct substances, although often allegedly conjoined. And the resulting onto-philosophy can be blamed as much on those who deny the very existence of one of these 'substances', usually the mental or spiritual one, as on those who assert the existence of both.[2] There are yet other examples of onto-philosophy that we have yet to meet amongst the postmodernists, incurred in their analysis of language, signs and texts. But more of this later.

It is sufficient for the moment to end this opening move in the present argument with the observation that imaginative investigation of and discourse about reality, and not just the abstract, analytic investigation and discovery about it, can result in its own characteristic manner in its own kind of onto-philosophy, or in this case perhaps onto-mythology of the merely imaginary. For imagination's *forte* is to envisage and describe how things interact in this unified universe, not only for continued well-being but for development, for betterment. In this creative reach of its remit the imaginative can easily end by positing the merely imaginary: powers or processes, things that do not at all exist as depicted, that are merely a matter of wishful thinking or of exaggerated fears. And what all of this amounts to in effect is the interim conclusion that both the imaginative and the abstract-analytical responses and approaches to reality, in so far as they can be adequately distinguished in any domain of human discourse, are equally capable of similar degrees of truth and falsehood in construing the real. And indeed each is as capable of correcting its own short-comings as it is of improving the other's performance. All of the great leaps forward even in the most physical of sciences – most recently relativity theory and quantum theory – are in origin leaps of imagination in construing anew the dynamic fabric of reality. Yet Einstein, for instance, could never quite bring himself to believe that the quantum picture of the physical world would turn out in the end to be anything other than provisional. Which simply goes to show that it is often difficult to determine whether or not one is in the pres-

---

2. Descartes himself was not guilty of the crude dualist metaphysics that is nowadays passed around under his name. Matter and spirit to him were 'incomplete substances' of the *une seule personne*. Spinoza represented Descartes's position much more accurately than most when he wrote, 'mind and body are one and the same individual conceived now under the attribute of thought, now under the attribute of extension.' (*The Ethics*, II. xxi. Note; see also *The Critique of Theological Reason*, pp 11ff.)

ence of a piece of onto-philosophy or onto-mythology or what is more usual, due to the mixed nature of human discourse, some hybrid of the two. Only the constant creative and critical engagement with reality can keep the truth a step ahead of the malign posse of conceptual and imaginative illusions.[3]

A more particular argument against the sole adequacy of an abstract analytic conceptualisation of *creatio ex nihilo* and its logic can also be raised upon an interrogation of the actual text of the opening chapters of Genesis. And this will lead to a challenge to the first assumption mentioned above, namely, that that text depicts a one-off act or activity that created the world, an act that constituted a space-time singularity as physicists would call it, by means of something resembling a peremptory command from the Omnipotent One.

'In the beginning of God's creating (*bara*) the heavens and the earth, the earth was without form and empty (*tohu w'bohu*) and darkness covered the face of the abyss.' The root meaning of the Hebrew *bara* is to cut, to cut out, hence to shape or form. The verb is used in Isaiah 43:1, 15 in order to depict Jahweh forming or fashioning a nation. At the other end of the Bible a more literal translation of Hebrews 11:3 gives us: 'the world (more literally still, the ages) has been framed by God's word, to the end that that which is seen be known to have arisen not from things which appear.' The key image for divine creation, then, is that of shaping, framing, forming and not of some act of irresistible force or power. *Tohu* connotes emptiness, the void; *bohu*'s connotation is more elusive: something unsubstantial, shall we say, something unreal; darkness is absence of light where no form can appear; the abyss the very symbol of formless chaos – a great collusion of images to evoke, with far more power and effect than an abstract and empty concept like nothingness could evoke, a formless and insubstantial unreality.

Add an understanding of temporal references that is characteristic of myth, and in particular of those grand metaphysical myths that deal in the ultimate origins or sources as well as in the final goals of all things: in the beginning, *en arche*, in the prin-

---

3. For an introduction to scientists like Gould who see science as essentially story, in opposition to postmodernists like Lyotard who see narrative as the very opposite to the experimental knowledge of science, see Stephen Prickett, *Narrative, Religion and Science*, Cambridge University Press 2003.

ciple itself; *in illo tempore*, once upon a time, long, long ago; at the end, in the consummation. No dates given or possible for, as Eliade so often says, this time of the myth is time out of time, a time contemporary with all calendar times, in short, eternity as the opposite of linear or even circular time. And the story you hear is a story of acts without sequence, of a sequence that belongs to the structure of the story and not to the (divine) activity as such – acts contemporary and collocated with every passed and passing point of space-time, as the cosmic myth always talks about the here and now.

So is the activity of the great cosmic/metaphysical myths contemporaneous with all points of empirical time. As pre-Christian Irish religion, for instance, illustrated by spreading versions of their creation myths and commemorations of what these connoted over the four quarterly rituals of each year, and then repeated them year after year. Then the creation of the formless, insubstantial unreality, the no-thing, is a contemporaneous by-product of the very forming in which the activity of divine creation essentially consists. This contemporaneous bringing-into-being of this formless no-thing as a by-product of the formation of finite things is quite comparable to (the origin of) the Greek image/idea of prime matter, that is, matter-without-form, which the Greeks actually called *to me on*, the no-thing. For only forms-of-matter are things, *ta onta*; and as that name implies, only already formed things can exist in their own right. Prime matter cannot be found on its own. None of this prevents the possibility of talking of created things being formed out of this no-thing; just as in physics measurable formations such as fields or particles can be said to emerge in or out of space-time, even though in reality space-time is the by-product of the emergence of such entities. Nor does any of it prevent the possibility of the no-thing being depicted as ever threatening the good divine creation.

This latter possibility is at least suggested by flood stories in which the abyss, the chaos returns to destroy a divine creation already compromised by evil. It is more than suggested in some Original Testament references to monsters of the deep, in myths as old as Genesis and as recent as *Moby Dick*. These are forces that, though created by God, albeit as by-products, are adversarial to good, forces that God can deal with but cannot in the activity of creation entirely avoid or simply eradicate, it would appear. And it is still more than suggested in Isaiah 45:7, a verse already quoted: 'forming light and creating darkness, making peace/prosperity (*shalom*) and creating evil, I am the Lord who does all these things.' The verb translated 'creating' here is the same word found in the creation story of the opening of Genesis, *bara*; and the Hebrew word translated 'evil' here is the same word found in the phrase about the tree of the knowledge of good and evil in Genesis 2:17 and 3:5.

So God forms all the things that make up the world, and forms them so as to make a universe, and in the process calls up unreality round the edges, as it were, of this reality. Just like darkness inevitably emerges at the limits of light, whenever a source of light sends out light. This darkness, this abyss, this formless no-thing-ness is sometimes depicted as engulfing or threatening to engulf the formed world, particularly in flood stories, for instance, that are themselves creation or re-creation stories. For the dry land has to be formed (separated) and furnished (once again) from the chaos/abyss in order that life should be possible. And this nothingness in turn may be interpreted either as an undifferentiated, that is, as Hegel argued at the outset of his *Phenomenology of Spirit*, an abstract nothingness in which we may imagine our world to be totally swallowed. Or, as Hegel would prefer to see it, it may be interpreted as a determinate nothingness. A determinate nothingness refers to the nothing-ing, the negation, the limiting in space-time, the deprivation of particular forms or stages in the course of on-going, advancing higher formation, so that higher states of existence and life are achieved. An example, in short, of what Hegel named the patient labour of the negative, in bringing all of reality to its final form and state.

The abstract concept of nothingness that characterises the most abstract analytic-conceptual version of the *creatio ex nihilo* theme, results most frequently in the commission of the sin of onto-philosophy. For that which is in actual fact an analytically abstracted element in a complex project – in this case no less than the whole concrete project of cosmic creation/evolving reality as we know it – is treated as if it were an entity, a thing in its own right. Thereby giving the very odd impression of 'nothing' then also as a something, out of which the world is created. An impression which then necessitates the corrections of confining the connotation of *creatio ex nihilo* to the two negatives outlined at the outset of this essay.

But how did this purely abstract conceptual version of *creatio ex nihilo* come about and take its place amongst the others? What follows offers one partial historical explanation. For the rest it is a matter of what Whitehead described as a retro-projection of the autocracy of the Roman Emperor onto the creator by Caesaro-papist authorities in the Christian church; or more generally a prejudicial-style replacement of the imaginative-mythic by the abstract analytic-conceptual.

The agencies directly involved in this creative forming of reality, according to the Genesis myth, are God's word ('God said') and God's spirit. The latter image is of the spirit as a great bird brooding over the abyss, a conventional image of forming in a womb and birthing or bringing into existence, in this case, a world. The former, however, the image of the word is more ambiguous. It could convey an image of word as an expletive, more especially a command, thus initiating a raw command scenario of creation, giving us the *Dieu Fabricateur*, as Sartre called him, who further issues with all thus fabricated entities instructions for use with equal power to command the obedience of any who might be tempted to use the creation for any purpose they choose – a total image of creation and creator much favoured by those who prefer their morality in the form of obedience to explicit divine commands mediated, naturally, through their own good offices. An image of creation and creator that finally yields images and ideas of divinely initiated yet humanly exercised autocracies. It is at this point of the development of ideas that the suggestion of imitation of Roman imperial autocracy comes into play, as an explanation of the manner in which Christians chose, out of all available connotations of creation out of nothing, the one which favoured the command-production of world out of a now abstract nothingness reified as a 'something'.

But the imagery of God's word could equally convey an intelligible form once again realised either in the thing and world formed by word, or in the representations of these in words and other signs and symbols, or in all of these. This latter option is offered particularly by the Greek word, *logos*, and was taken up explicitly, even if it too was sometimes bowdlerised in a number of ways by influential Christian theologians. But in spite of all of that, a further brief visit to the pre-Christian origins of a philosophical theology that formed almost the whole of the theology which Christians borrowed and adapted to their own message would be enough to convince any student of *creatio ex nihilo* of the variety of its versions and interpretations. And of the age, persistence and variety of the forms of myths or the forms of abstract analytic conceptualisations, and especially the forms of philosophical commentaries on Plato's myths, in which the image/idea of creation out of nothing was available for borrowing.[4]

So from at least the Middle Platonists (roughly over three centuries stretching from BC to early third century AD) this philosophising about creation often took the form of commentary on Plato's inspired scriptures and in particular on the story (myth) of the *demiourgos* (the divine creator of the universe) in the *Timaeus* (esp 28ff). This story was of a divine creator, a mind or *logos*-type entity, forming a world by means of Platonic Forms now in the Middle-Platonic period seen as forms (or *logoi*) in the mind of the divine *Logos*. Forming world in or out of what? In imaginative terms, a receptacle, a space, a disorder, an emptiness of forms and of their ensuing order. In more conceptual terms, 'material' or 'matter', these words in parentheses because they do not wish to refer to formed matter, matter as we know it, for that is already formed, created. They refer rather to something like the 'prime matter' of Aristotle's hylemorphic theory, which never exists on its own like a formed thing, a *physis* or nature.

And then some commentators characteristically debate whether Plato's myth implied a creation in time or not. Those who say, yes, inevitably implying that there was matter, a resistant, formed and evil force awaiting the *demiourgos* and its best efforts. Examples of this would then be those Gnostic dualists so

---

4. Many other terms for images/ideas of origins such as emanating from, taking origin from, being brought forth, being formed by, and so on, are fully interchangeable with 'being created by', or 'made by'. Not least in that all of these, receiving as they do their humanly accessible connotations from such actions/experiences as are characteristic of creatures, need to be treated to the 'ways' of negation and eminence simultaneously with the 'way' of affirmation. As the Greeks also insisted (see the *Didaskalikos* of Albinus), once again before Christians began to philosophise their faith, that is to say, to theologise, and that not just in imitation of the Greeks, but by borrowing the theology of the Greeks wholesale. In any case simplistic and misleading contrasts between emanation and creation should not be allowed to lessen our true appreciation of the amount of borrowing in which our ancestor- theologians engaged.

sternly and lengthily denounced by Plotinus, for instance, in *Enneads* II, 9. Those who say, not, implying that the talk of a beginning referred to a necessity of the story form, but that in reality the world we know came, not from a moment in linear time, but from eternity. In that case, that which is depicted as chaotic flux, receptacle-space, prime matter, darkness (in which no form appears), comes into existence with the creative out-pouring of the good creator (as Plotinus puts it in *Enneads* IV, 3.9; III, 9.3). Just as Einstein would say that space is a function of the forming of 'fields' rather than a receptacle which pre-exists or exists independently of them, and parts of which may still lie empty of them. Yet it is still possible to think, imagine and say that the world-forming activity takes place in or out of this prime matter, this no-thing which is symbolised by images of formless, chaotic abyss, and of darkness; *creatio ex nihilo*.[5]

And it is also possible to say that this by-product of divine creation then acts as adversary to the creator and all creatures, for with the exception of the raw command model of divine creation, and the abstract analytic versions of it in which that model finds a home, the *creatio ex nihilo* theme does not entail any of those extreme conceptions of divine omnipotence as autocracy so objectionable to modern theologians, and particularly to feminists, and thought to be the source of divine autocracy then claimed by Christ's vicar on earth, and so on. Quite to the contrary, those early creation-out-of-nothing contexts that make such good combined use of imagery and idea, always make clear the limits of what even a creator God can do. This is blindingly clear in the *Timaeus*, where it is plainly claimed only that the creator makes everything good and not bad, 'in so far as this

---

5. When in the spoken version of this piece I referred to Plotinus and Porphyry, and in particular to the latter's commentary on the *Timaeus* for examples of the creation-out-of-nothing theme, I was taking these as the best examples of this theology amongst the Greeks. I did not feel it necessary to illustrate how in this respect they were Neo-Platonic exponents of a long tradition that went back through Middle-Platonism. Those who would still wish to assure themselves of this long tail-back should read Philo, a contemporary of Jesus, a Middle-Platonist in philosophical terms and a thoroughly orthodox Jew who, particularly in his *On the Creation of the World*, offers this kind of version of creation-out-of-nothing, but now by exegeting also his favourite creation myth, from Genesis. And it is well known that his future influence was amongst early Christian theologians, rather than amongst his fellow Jews.

was attainable /' in face of the resistance of something called 'necessity'.

Translate that however one likes into ideas of 'the materiality of things' inevitably entailing the finiteness of space-time and the consequence of flourishing at each other's expense. Follow this with its implications of the inevitably of the disintegration and death of such gross material bodies. Follow that in turn with its usual accompaniment of loss, grief and loneliness. Follow all of this further with that general fear of disintegration and death that makes one grasp at finiteness to sustain oneself, as Kierkegaard once put it, with ensuing damage both to finite things, especially persons who cannot bear such burdening expectations upon them. And one can all too easily understand how some earthly powers, instead of co-creating as they are designed to do, freely turn destructive instead, if only in the process of accumulating earthly possessions in a vain attempt to postpone death, because they are created free to do as they decide. And from all of that, one just begins to get a sense of the loving impotence of a God who formed forms of existence to form in turn all that would evolve in this wonderful and potentially ever more wonderful world. A God who cannot avoid being the ultimate source, both directly and indirectly, of evils natural and moral, and can only, and only with the co-operation of creatures, keep on creating out of what Plato called the Creator's own overflowing goodness, *Eros*, love. Omnipotence, then, need not and does not connote absolute autocracy, but rather the ability to create all that is created: Father Almighty means 'creator of heaven and earth', that is to say, of all things, 'seen and unseen'.

The Stoicised Platonism, then, that Christians of that early period borrowed to form the substance of their own theology, envisaged the divinity called *Logos* continually forming and reforming the universe in and through the derivative 'seed words', the *spermatikoi logoi* or *rationes seminales*. Hence the divine *Logos* by whom all things are made the Stoics also named as *Physis* or Nature, after Aristotle's definition of *physis* as the form of things that have in themselves the source of their own motion or development. This was, and is, a story of continuous creation which is quite compatible with the Genesis account of God creating things containing the seeds of their own future within themselves; and creating one species in particular which, being in the image of the creator, knew (named) the creatures and was therefore placed in a position of co-responsibility for the future of all. This idea of the human species as the husbandman of creation, the form of creature which was informed of itself and of the others and hence could continue to guide the continuing formation of all, translated quite easily into the Stoic system in which the characteristic *spermatikos logos* of the human was conscious both of itself and, potentially, of all other *spermatikoi logoi* – and was therefore in a position to participate more fully than the others in the nature and work of the divine creator *Logos*. Or, to put the matter the other way round, to have the divine *Logos* work its continuous creation in and through the co-operation of the human *logos*.

Put together a number of these thoughts about the Genesis myth of creation and its more philosophical exposition from the beginnings of Christian theology – the thought that what the myth describes is what happens at all times; the thought that the act of divine creation is an act of forming rather than an act of raw power which on the analogy of a command brings a thing or state of affairs into existence instantly and fully formed; the thought that the divine former of things and of the relationships between them that make them a universe, works through these forms in order to their fuller or future development; and the thought that, due to the materiality, plurality, hence finitude of things created, which is itself an inevitable by-product even of divine creation, the life of one, and especially advance in life, is at the expense of some part of one's own life and also, most likely, at the expense of the lives of others, whether the ones and the others be individuals or species – and you have the outline of a Christian concept of divine creation fully compatible with the contemporary concept of evolution. Evolution is then the visible trace, seen from the worm's-eye view of the creature, of the universal continuously creative power that drives the formation and continual re-formation of this ever-changing universe, through the mutual information, transformation and, yes, deformation that the created forms exercise upon each other. Creative evolution; evolutionary creation.

That is not to say, of course, that the post-Darwinian concept of evolution can be found in Genesis or even in the Stoicised Platonism adopted by early Christians as their philosophical exposition of Genesis. But it does allow us to take contemporary evolution theory fully into our Christian doctrine of creation, if only as the best account available to date concerning that inherent developmental capacity which for Aristotle characterised *physis*, the natural form(s) of the creation, and which both Genesis and the Stoic philosophy of divine creation presented under the image of the seeds. Although the Stoic philosophy saw this development to involve much more than the Genesis idea of reproduction, and much less than the vast complexity and range of development that modern evolutionary science opens up to our still somewhat astonished gaze.[6]

Two things need to be said concerning the choice and deployment of dynamic form and forming as that which reality reveals to be the master-image for continuous creation, the sheer cosmic dimensions of which suggest traces of the divine. First, a word about mind and matter again, and how some ultimate

---

6. This creative-evolution model of the divine creation of the universe has existed from time immemorial, and in contexts far different from certain readings of Genesis or the Platonised Stoicism of early Christian theology. But it was kept out of the mainstream, as is so often the case with aspects of the Christian truth that were lost to the sight of those who, in pursuit of tight orthodox definitions of the faith, adhered ever more closely to a model of command by omnipotence. So from Ovid's *Metamorphoses* to the *Mahabharata*, and to pre-Christian Irish shape-changers, the theme of divine power within, taking the shapes of and changing the shapes of different creatures, has been universal. And with the advent of modern evolutionary theory, metamorphosis of the whole of natural reality is secured (Marina Warner, *Fantastic Metamorphoses, Other Worlds,* Oxford University Press 2002). A related arena for the survival of this more rounded idea of divine creation as continued exercise of immanent creative evolutionary power – 'the force that through the green fuse drives the flower,' as Dylan Thomas described it – consists in that esoteric European tradition rather disparagingly known as The Occult. Deriving from a variety of ancient sources, instanced by Heracleitus, Pythagoras, Neo-Platonism (including the great Irish philosopher, John Scotus Eriugena's *Periphysion*), the adherents of this persisting movement, from Jewish Kabbalah to the Christian William Blake, sought by means conative as much as noetic to achieve a unity with the divine creative impulse throbbing not only through their own souls, but through all the changing forms of the whole universe, and thus derived a religion of creative co-operation and of an ethic uninterested in merit (B. J. Gibbons, *Spirituality and the Occult,* London: Routledge 2001; Shimon Shokek, *Kabbalah and the Art of Being,* London: Routledge 2001).

source of space-time and of all that evolves in it, while conceived as of the nature of mind or spirit, could give rise to the evolving forms of space-time without these incurring the suspicion of being simply carved from its own *ousia* (a Greek term translatable as either being or substance). And, second, a word about transcendence and immanence.

Sartre objected to the Christian idea of a divine mind creating the universe on many grounds, but first and foremost because he maintained that no conceivable act of such an entity – if it existed – could ever result in anything that could then be deemed to have broken out of its own absolute subjectivity. Now, of course, it is easy to retort: well, he would say that, wouldn't he. For his own philosophy is both exponent and victim of a completely dichotomous dualism of *pour-soi* and *en-soi*. And this in turn is the complete counterpart of that equally dichotomous dualism of two substances falsely attributed to Descartes, though Sartre himself preferred to talk of two dichotomously distinct realms of reality. But it is more important for present purposes to realise that Sartre is just one of a group of philosophers and assorted scientists of the modern and postmodern eras who commit the sin of onto-philosophy in their dealing with the 'substances' of mind and matter. And that many of them then go on to assume that the *res cogitans* of 'Cartesian' dualism is just another name for a material brain and nervous system, so that there really is only one substance, the *res extensa*. Or else they struggle to see how mind or spirit or consciousness and resulting activities thereof could derive from the known evolution of the purely material, extended substance – with so little success that John Searle sighs for a second Newton.

So immersed are those moderns in their version of the sin of onto-philosophy, so biased towards the resulting crudity of their materialism, that it could possibly never occur to them to consider how the *res extensa* could derive from mind or mind-like being. And this, paradox of paradoxes, despite the fact that Kant, as crude a dualist in his way as the modern age has produced, with his distinction of noumenon and phenomenon, yet equivalently declares at one critical point of his first great *Critique*, that the defining features of matter, of material substance, namely, its extension in space and time, are known by us to derive from mind. For space and time, he said, were *a priori* forms attributable to mind in the process of sense experience.

Einstein thought that Kant could hardly be taken seriously in his suggestion that space-time could be a purely subjective experience. Yet Einstein also insisted that, first, matter has lost its role as a fundamental concept in physics, as the model of field proves more fundamental than that of particle or material point. By carrying this replacement to its logical conclusion, space-time becomes a structural quality of a field. Now the idea of a field – the representative of reality, as Einstein calls it – a field of energy, let us say, expressed in mathematical form or formulation, is much more of the nature of a mind-like entity, or at the very least something formed or fashioned by mind-like entities. Certainly, with this demotion of the crude idea of matter that was characteristic of the equally crude dualism commonly called 'Cartesian', the prospect of a world as we know it deriving from mind, becomes much more promising for philosophical investigation. And much, much more promising than the prospect of mind deriving from the cruder conceptions of matter. So, however one tosses the penny, the crude dualism called 'Cartesian', a consequence as it is of the sin of onto-philosophy, is to be rejected in favour of a much more subtle metaphysic of mind-in-matter and matter-in-mind.[7]

Such mutual in-being of mind and matter brings us to immanence, and to our second and final word in this section, concerning immanence and transcendence. Although both of these terms have a basic connotation construed from spatio-temporal imagery – remaining within or outside of; going or being borne beyond – their metaphorical use, which is the usage at issue in the kind of contexts we consider at this conference, does not need to carry with it the mutual opposition between immanence and transcendence inevitably suggested by the most literal, that is to say spatial connotation of that imagery. For according to that most basic connotation, to the extent that something remains within, to that precise extent it is not (gone) outside or beyond; immanence is at the expense of transcendence, and vice-versa. But when these terms are used metaphorically, in a usage

7. See Appendix V in the 15th edition of Albert Einstein's *Relativity: the special and the general theory*, London: University Paperbacks 1954. For a fascinating piece on Spinoza and Weil which incidentally illustrates the actual equivalence of saying that the natural world is outside the mind and that it is inside the mind – and vice-versa, presumably – see David Cockburn's 'Self, World and God in Spinoza and Weil,' *Studies in World Christianity*, IV. 2 (1998), 173-186.

that carries them beyond our more primitive experiences of the spatio-temporal relationships of bodies, such mutual exclusion need not at all apply. Certainly this is the case when they are used of a creative power that forms a world of things, as form-ings of the forms of space and time. For this is a creative power that makes these forms inter-related and interdependent to the point of forming universe. So that they are taken up into the con-tinuous creativity of the universe, in that each species-form through its individual members is empowered to inform, re-form or deform, and in any case to transform both itself and the others. Then, certainly, such a foundational creative power, whether it be named a gene or a god, can be said to be at all times and places as immanent in space-time as it continually transcends each stage in the evolution of the same universe.

Yet so much talk of immanence and transcendence sounds as if it takes immanence and transcendence, terms for abstract con-cepts that connote two concomitant and intrinsically inter-related features of reality-in-the-process-of-becoming, and reifies them. Treats them, that is to say, as concepts for separately existing en-tities or separate and distinct regions of being: one within space-time, the immanent, and the other beyond or outside of space-time (whatever 'beyond' or 'outside' could mean in this con-text), the transcendent. This is surely another form of the sin of onto-philosophy.

The anthropologist Levy-Strauss, who thought of cultures as texts, once famously remarked that incest is bad grammar. And one cannot help thinking that a little lesson in grammar would prevent much of the current talk about immanence and tran-scendence from misleading philosophical discussion in the way it too often does. So here now is an example of the kind of lesson in grammar that might serve to correct the main misuses of these twin terms. 'Transcendence' is an abstract noun. It connotes a process in which a variety of entities are known to engage, a process of over-reaching or going beyond certain current limits or limitations. 'Transcendent' is an adjective for entities as they engage in this process and, as with any adjective, the noun it qualifies should be added or clearly implicit in the discourse. In short, whenever the phrase 'the transcendent' is used, it should be accompanied by a noun which answers the question: the transcendent what? The transcendent mind? The transcendent gene? The transcendent process of natural selection? Use of the noun 'transcendence' or of the adjective 'transcendent' with nothing but the definite article in attendance could be expected to give rise to much mystifying gobbledegook. And wherever academics are gathered in a conference designed to discuss transcendence and/or immanence, that expectation is seldom disappointed. Especially in a context in which the expectation seems to be that transcendence or the transcendent may be taken automatically to denote divinity. Even if no reason has been supplied for the existence or character of any entity or region of being that could reasonably be deemed divine. Therefore, far too much talk about transcendence or the transcendent in fact provides instances of onto-philosophy that occur in the precise form of that onto-theology so frequently complained about (in others, of course) in modernist and postmodernist texts.

Creation, in all the most obvious instances of it, involves a process of transcending in that it brings about new forms of things or of the processes in which things engage and from which they result. And if people believe they can detect within all this pullulating and pulsating universe a unified and unifying agency that forms space-time and forms (or by forming) the other natures that take up or make up space-time, continually combining them so that they participate in that perpetual transformation seen from this side as evolution, then it may well be claimed that people are in the presence of the traces of a being which, as the constant creative source of cosmic or universal reality, can be deemed divine. Transcendence and immanence then are not contraries but co-ordinates used to locate these agencies that, in their mutual co-inherence, bring about those states of disrupting, exceeding and so on which David Wood so well describes.

Such are the end results of analysing what we call creation through the master-image of creating as forming or shaping. And there are similarly interesting results for the understanding of the relationship between the creator and the creatures. These may best be seen from a brief revisit to the contrast with the other master-image of creation already mentioned: the command-of-an-irresistible-power image, more crudely known as the shouting-it-into-existence model. On that latter model, God would need to shout again to tell us about it who were not there to hear the first shout. And then She would need to shout again – more than once, given our persistent waywardness – to tell us how to behave ourselves in and with the creation – thus yielding the model of divine revelation as verbal communication and the model of morals beloved of Pope John Paul II, as a code of commands requiring as the first moral virtue a child-like obedience. Further, the command model of revelation (telling us what to believe), for morals (telling us how to behave), and for existence itself, suggests a creator God who operates from outside the creation. Whereas the forming and re-forming model of creation used from Genesis onward opens onto the model of the *arche anarchos*, the beginning in the sense of origin or source-without-source, ever creatively, transcendentally active within the evolving cosmos. This yields a model of knowing the divine through the traces of this continuous creativity, and more precisely through the experience of being oneself both passive and active with respect to this cosmic creative evolution. Essentially a pragmatist model of revelation and knowledge, informing by creative forming. And a model of morality as co-operative envisaging and attempting better being, more fulfilled life and existence for all inevitably interlinked creative creatures; real responsibility, with all the damage and all the guilt, yet all the fulfilment and all the dignity this can bring.[8]

II. POSTMODERNISM

It is best to begin this section also with some common assumptions, and mainly once again, with critical intent. It is commonly assumed, even when it is not being trumpeted by prominent postmodernists themselves, that postmodernism dispenses with meta-narratives, those grand narratives about being which attempt to tell the truth, the whole truth, and nothing but the truth of 'whatever is begotten, born and dies ... of what is past, or passing, or to come,' as Yeats would put it. Or, in another formulation of the same assumption, that postmodernism finally dispenses with metaphysics. The first formulation of the assump-

---

8. It is hardly necessary to produce and itemise a full list of instances in which the Bible alternates images of immanence from God-in-creatures to creatures-in-God: 'in him we live and move and have our being,' 'I live now, not I, but Christ lives in me.' And it is interesting to note that the same would apply to Dawkins' selfish gene, which some of his critics see as a creative agent modelled upon a post-Enlightenment *Logos*-type creator divinity (see Steven Rose, *Lifelines: Biology, Freedom, Determinism*, London: Penguin 1977).

tion (and I think it is the same assumption), is best dismissed by a simple *tu quoque*; although it would take another book by Sean Burke to do it properly. Burke's book, *The Death and Return of the Author*, the best book on postmodernism by one who does not profess philosophy, shows how the author also ushered out of the front door by postmodernists, always surreptitiously re-enters their scenarios by the back door and operates all the more effectively for not being any longer noticed. This is an author incidentally who, when noticed, looks suspiciously like these divinities in the old meta-narratives who, sitting outside the world and subject to none of its changing conditions, eternally and infallibly and immutably know, quite literally, all about it. A companion volume by Burke could make a similar case concerning the much advertised demise and surreptitious return of a very similar kind of meta-narrative.

A similar critique of the premature advertisement in post-modernist publications of similar assumptions of the demise of metaphysics, can bring the important philosophical matters at issue here under an even closer scrutiny. But what do we mean by metaphysics? In his work of that title, Aristotle is engaged in what he himself called *prota philosophia*, first philosophy, and in this he included both the study of beings precisely as being, and the study of that which could be found to be the foundation, cause or source of this being of all beings. And all of this duly entered into the connotation of metaphysics in the tradition of Western philosophy. Heidegger, however, preferred the term 'ontology' for the study of beings in respect, not now of their various particular forms or natures, but of the condition of their being as such. But for that part of traditional metaphysics that studied the supreme being alleged to found the very being of the beings, he used the term onto-theology.

For, as Heidegger argues at the outset of his *Was ist Metaphysik?*, those who think they have sufficient reason for believing in the existence of a supreme being which founds and thus guarantees the being of all empirical beings, cannot even engage with the very question which the being of all such empirical beings puts to them, and which yields the whole substance of the discipline of ontology or metaphysics. The question, namely: since the being of all space-time entities seems to be a being-unto-death, a being-unto-nothingness, 'why are there essents, why is there anything at all rather than nothing?' Those for whom an answer to that question is already available – whether through rational proof or special divine revelation – cannot experience as a real question that question that is at the heart of all finite existential experience. It can only be for them an 'as if' question, and the pursuit of it a pure philosophical charade. And the onto-theological tag comes into play because these so-called proofs and their results are simply instances of an abstract conceptual analysis and logical argument, the terms of which are then taken to represent realities defined by the analytic and argumentative process itself.

So for instance, abstract concepts of omnipotence, omniscience and so on (all the omni's) are assembled, with ideas of personality and of 'that than which nothing greater can be thought' thrown in for good measure (Kant was correct is suggesting that the so-called Ontological Argument was actually assumed in all of the so-called cosmological proofs, though he failed to notice that it would have to be equally assumed in his postulating a god on moral grounds). Abstract concepts of contingency and necessity are then added to the mix in order to provide what seem like logical steps in an argument. These abstract concepts are then substantivised, and of course capitalised as The Omnipotent, The Omniscient, The Necessary Being (and of course, The Transcendent), *et voila!*

Something similar to Heidegger's strictures upon any return to metaphysics happens again in recent reaches of that long-tailed British, that is to say, Humean scepticism that has for so long masqueraded as empiricism. At first ousted as plain nonsense by the non-sensical over-simplicity of A. J. Ayer, his logical positivist predecessors and linguistic analyst successors, metaphysics has recently made a come-back to these otherwise rather barren philosophical regions. But, first, the come-back was condoned only under such straitened conditions as those imposed by the philosophers Quine and Strawson, who insisted that metaphysics should confine itself to a critical analysis of the most universal existential structures of our empirical universe, as these seem to emerge at the frontiers of the advance of the physical sciences. Metaphysics must never again pretend to an access to a knowledge of the most general conditions and structures of existence that can somehow by-pass such persistent scientific attention to all that can be seen of the world from within.

And yet, second, it seems as if British metaphysics has since gone beyond the strictures imposed by Quine. It has gone further into the deep existential structures of the world and the vistas of possibility there allegedly opened up, further than modern physics would countenance. Or has it, really? Some contemporary physicists who talk in terms of string theory, as will be pointed out in the *Epilogue*, have gone further into the metaphysical realm, not least in talking of realms of possibility, than Quine or Strawson or many of their more sober scientific fellows would countenance. But, however that may be, neither British metaphysics nor the science it takes as its stamping ground, will yet agree in general to even envisage the existence of a supreme being, source and ground of all the other beings in the world and of its deepest structures.

What is noticeable about these various versions of re-emergent metaphysics is this: the supreme being envisaged, if only for purposes of rejection, is always envisaged in the image of one that squats outside the world – an image used by Heidegger but which goes back to the Marxist-Feuerbachian critique of religion. An image of a supreme being that fits with Strawson's rejected view from outside and that is often adopted for purposes of defining a revisionist as opposed to a descriptive metaphysic. An image of a supreme being that is reminiscent of the 'shouting it into existence' imagery of one reading of the Genesis myth of creation by word, and reminiscent too of crude though common understandings of transcendence as also referring to something outside rather than within the entity or event with respect to which the process of transcendence is deployed.

Now there is no reason for saying that this outsider status for the creator was invented by those people who criticised it. Quite to the contrary. In the West at least it was suggested by Christianity's insistence on a kind of faith in God which itself placed God beyond the range of investigative reason, and hence beyond the range of Strawson's descriptive metaphysics. And even though certain Catholic traditions in Christianity seemed to place God's existence well within the range of investigative reason, they did this mainly by adopting what became known as proofs of God's existence. Then, particularly since the dawn of the Age of Reason or, rather, the Age of Rationalism, philosophy of religion took these odd pieces of reasoning to be its first and main pre-occupation.

In the light of this contemporary status of metaphysics then, what is now the relationship of postmodernism to metaphysics? Well, first of all, the postmodernists and their followers chant the mantra of the overcoming, or of the end of metaphysics as ritually as any others. Yet Sartre was quite correct in observing that any metaphysics implies a theory of knowledge, an episte-mology, that is, a theory of how reality is known, and then of how is the knower known. Just as any theory of knowledge, even in the form of theory of text or language, implies a meta-physic, that is, a theory about the characteristics and conditions of being as such. A theory that seeks to go beyond the more par-ticular theories that deal with specific kinds or forms of reality which Aristotle called *physeis*, natures. These are particular forms of being that contain within themselves the source of their development (so kinds of particles, for instance, or species). So that the effort to describe and explain the characteristics or con-ditions of the whole of empirical reality, characteristics and con-ditions common to all natures, all *physeis*, *le tout ensemble*, could be called *meta ta physika* (that which went beyond the individual natures, species and classes), metaphysics.

And in both cases, that of the epistemology and of the meta-physics necessarily implied from the beginnings of the scientific quest for knowledge of the fabric of reality, it is philosophy's task to make the implicit critically explicit. Quine says no more and no less when he insists that hard science, in order to secure the truth of its best theories, needs both an ontology and an epis-temology, and that it is philosophy's task to do its best to supply these. That Sartre and Quine are correct in these views – together with so many others who take all this talk about the end of meta-physics to be about as effective as whistling in the wind – can be shown by reflecting for a moment on a deep and serious fault-line that runs through Derrida's meta-narrative.

This faultline in Derrida's thought can be best described as a serious metaphysical deficit; and it can be most easily detected by closer inspection of that Saussurian semiology which Derrida imports into his system; and more particularly the Saussurian definition of sign. The sign is defined by Saussure as the associa-tive totality of signifier and signified. Where the signifier is some image of the psychic order, acoustic perhaps or visual, a word heard or read, and the signified is something of the order of an idea. The signifier itself is empty; only the associative totality of itself and the signified has and gives a meaning. And from here Derrida takes off into his theory of differance, invoking simult- aneously the ideas of differing and deferring. For signifiers are never anchored by the signified, but only by other signifiers, by their very differences, and so any final meaning or truth for any of them is infinitely deferred; and so on.

Now unless one's critical faculty has fallen asleep, one will feel at this point that something must immediately be said about sign so defined. And it is this: signs exist only as part of a well- known and quite distinctive actual, real existential process and one which, incidentally, is by no means confined to human be- ings. The real, existential process in which alone signs exist, in which alone signs are and are signs, is a process in which one communicates with another about something, even if that some- thing be only the one or the other communicator, or only things with which the one and the other have to do. Outside of that ac- tual, active, living process, marks on paper, sounds in the air, images and ideas of a mind, or any and all combinations of these, are not signs, or parts of signs, or anything to do with signs. As Wittgenstein observed, there is and can be no such thing as a pri- vate language. Or, to put the same point more positively, lang- uage in order to be and to be language, whether in textual, oral or any other form, must be public, that is to say, instantly usable in the most public forum. And in order to be public it must be both objective and subjective or, rather, inter-subjective. That is to say, it must denote and connote objects in the world that are available to at least two consciousnesses-in-communication.

In order to be persuaded of all of this it is necessary only to explicate a little further the process by which signs are and are signs, the process by which languages, being sign systems, come into existence. As the ancient Latin poet put it, with that awe- some power of precision that poetry alone possesses, *sunt lacrimae rerum, et mentem mortalia tangunt*: tears are of things, and mortal things do touch the mind. For 'tears' say 'signs' for tears too are signs, and then what is being said is this: things are intel- ligible forms (*logoi*) of matter, hence finite and mortal; and being such they inform minds, and minds then know and can judge them. In this case of the knowing of things our poet intimates that things are known through the emotions they incite. For in- deed the emotions, as well as imagination and intellect with which they are so closely aligned, are heuristic devices in their own right. The Stoics were wise enough to know and say that our emotions are judgements, primary means of apprehending what objective reality reveals, and of communicating this to others. In this case of the poet's imaging-forth, the metaphysical condition of the mortality of things material touches the mind. It is apprehended in our sadness and communicated in our tears. All part and parcel of the commonest semiological currency of our kind.

So the intelligible forms, the in*form*ations that come from reality in its constant self-revelation, in the course of our constant interaction with it, are then communicated in signs. In a vast variety of these, ranging from signals that simply draw attention to something, accompanied perhaps by expressions of emotion that interpret it (as fear does in pointing at a predator), through pictorial representations (cave paintings at Lascaux), or the special sign that is known as a symbol because it participates in the reality it signifies and is thus frequently performative (the Christian eucharist where bread and wine, both gifts of nature and human products, are broken and poured to others, to show that we live by giving life, that is, by dying for each other, as life and its supports are constantly poured out unstintingly to each and all from their original source), to the more arbitrary, conventional kinds of signs that form these extremely complex examples of languages that all peoples evolve, comprised of words spoken or written, or gestures (as in so-called sign languages) and themselves ranging from most concretely imaginative ('radio has the best pictures') to the most analytically abstract ('Postmodernism: Transcendence and Beyond').

Perhaps that is to go on a bit much about such an elementary matter, but it never fails to be surprising how difficult it seems to be to get certain postmodernists and their followers (and indeed assorted phenomenologists also) to see what appears to be so obvious. Signs exist and can exist only in the existential context of inter-subjective communication concerning *ta onta*, the things that are. Leave out of the account of signs then either or both of the following: the things that in their self-revelation offer the first forms from which signs may be developed, and the subjects who can both develop sign systems from these and communicate them to each other, and the loss suffered comprises both signs themselves, and since texts are made up of signs, the loss also of even a text. And this last is a loss for which no amount of chatter about some mysterious ur-text can in the least compensate. So that all that remains after all that loss is a prime piece of onto-philosophy which treats an analytic abstraction from the full descriptions of signs – the associative totality of signifier and signified (like sound and fury) signifying nothing beyond itself – as if it could exist in its own right, when in fact it does not and cannot.

The loss of subjects and things from this account of reality, the loss of their informative interactions amongst themselves and with each other, as the existential context in which alone signs are and can be signs – that, and nothing less, is the enormous extent of the metaphysical deficit in Derrida's philosophy. It is a metaphysical deficit which he himself seemed to acknowledge – though still not fully and completely – in that most pretentious and oft-quoted declaration of his: *il n'y a pas d'hors texte*, there is not anything outside of or beyond text. And despite the number of times in which he and his expositors have tried to re-interpret that statement so as to prevent its plain connotation from coming across, all such moves are futile, for the metaphysical deficit is entailed by the very importation into the heart of his system of the fatally flawed Saussurian definition of sign.[9]

It might be possible to suggest that the ur-text to which Derrida refers is comprised in fact of that inherently evolutionary reality that continually comes into being precisely by the natural forms (idea-like entities) that constitute the natural cosmos by forever informing, reforming, deforming, transforming themselves and each other, in an open-ended, infinite process. So that the very truth we tell about this evolutionary universe – as indeed whatever truth we may be able to tell about any ultimate source which we can only see and describe in terms of our

---

9. I realise that in my original oral presentation of this piece I relied too much on that (in)famous declaration of Derrida's, *il n'y a pas d'hors texte*, in order to make the case concerning the metaphysical deficit in Derrida. It is always possible to try to wriggle out of the plain meaning of one's words, by continually insisting that one did not mean *that*, and then go on to say what one did mean (but did not quite say?) And indeed in the following paragraph of the paper I look at a possible meaning for the declaration which might allow it an acceptable meaning – but at a cost that Derrida might not wish to pay.

present images and ideas about that universe[10] – is consequently subject to continual revision. But if that is the case, it would still be necessary to say that Derrida's system is more of a caricature than an accurate picture of such a cosmic text. A caricature exaggerates in order to draw attention to some facets of the thing depicted, but this caricature does this to the point where other equally important facets of the reality in question disappear entirely from view. The facet which appears in Derrida's picture of reality exaggerates the mental nature of the reality portrayed: signifiers and signified both as conscious, mind-like creations. The facet which does not appear at all in his picture – thereby undermining the value of a caricature by an equal and opposing falsification – is the process in which this multi-formed, ever-transforming reality is expressed and emerges in the actual, material, existential conditions of space-time, in which other minds can contemplate, as well as to some extent re-create it in both mental and extra-mental terms.[11]

But Barthes's epistemology and accompanying metaphysics is not so inadequate. Of course he also uses Saussure's inherently inadequate and indeed fatally flawed semiology in order to

---

10. Joseph O'Leary argues most convincingly in 'Ultimacy and Conventionality in Religious Experience' (as yet unpublished) that although the Ultimate is referred to and comes to be known by us as simply Emptiness, Unconditionedness, Un-formed, No-thing-ness, and so on, our knowledge of it is still necessarily characterised by the conditions or forms which we perceive it to transcend. As he puts the point with reference to Madhyamika thought: 'emptiness' is always 'emptiness of'; ultimate truth has as its basis some conventional truth; the unconditioned dawns on a conditioned mind, and emerges as the dissolution of just those conditions already in place ... in practice emptiness emerges on each occasion as a deconstruction of a given construction. Not the endless deconstruction of Derrida's differance, to be sure, but a deconstruction that finds something unconstructed, unconditioned at the heart of the constructed, conditioned. Ultimacy is always known as a conventionality deconstructed.' Apply this to Christian Neo-Platonic quests of the Ultimate, as indeed to all other religious traditions, and it is easy to see that knowledge of the Ultimate must develop and differ as one moves forward within each tradition in one's knowledge of the conditioned, formed universe; as also if one moves between these traditions.
11. A good illustration of what is meant here can be found in Richard Southwood's recent book, *The Story of Life* (Oxford University Press 2003). The book recounts the steps through which the secret of life's

rightly rid the world of those onto-theological systems whose conceit it is to describe reality in immutable and obligatory categories allegedly available in or through some being squatting outside the world – and that, incidentally, is just about all that Sausserian semiology is good for. However Barthes then, without giving any notice of this, moves to what is in fact an adequate semiology, with accompanying epistemology and metaphysics.

Barthes's semiology-epistemology and (at the very least, implicit) metaphysics is essentially comprised of two axioms. The first, derived from Marx, states that the world we know bears everywhere the traces of humanity's intentional and active presence in it.[12] As Barthes puts it: the most apparently 'natural' thing in the world retains a trace, however faint in some instances, of a presence of human being and of human acts of producing, managing, using, subsuming or even rejecting those things of which the natural world is constructed. The second

---

origins and persistence is explained: Darwin's primordial soup; inorganic molecules leading to complex carbon molecules, to 'protocells' and then to living organisms; organic molecules spontaneously emerging from elements already around in the early history of the earth; and, finally, the model of DNA, an information system complete with strategies for at once transmitting and altering itself. At that last stage many felt life's secrets were finally in the open, until some more critical minds realised that progressive theories in chemistry showed only the raw material in and through which the process called life occurred. And that in Southwood's view still left us with one large question to answer: how do organisms know how to form and re-form the information they pass on? Clearly in posing this question he rightly considers the organisms to be knowers. But that means that further investigations need to be made. Does each have to know how to organise itself and others so as to keep the whole on track in that process of continuous creation known as evolution? Or is there present in this process some cosmic knower organising and transforming the whole through all individual organisers, transcending all from within? And how is that knower to be known? Derrida's epistemology-metaphysics is clearly inadequate to the task outlined in such scientific-cosmological quests for knowledge.
12. The world is humanity's extended body according to Marx. Which prompts one to suggest that some Christian ecologists who have pressed into service the idea that the world is God's body, should perhaps produce the metaphysics which would at once clarify and justify that concept, which one can perhaps see that they need, but not quite see their title to it.

axiom states that language in essence is designed to do things: *dresse ... a les agir*. One recalls Heidegger's word for things, the old Greek term, *pragmata*, things with which we have to do. Put the two axioms together and the resulting epistemology-meta-physics is as follows: we know things, and each other, in the process of acting together on and with them and being acted on by them. Knowledge is our consciousness of the forms of selves and things as these are informed, reformed, deformed and in all cases transformed in that ubiquitous and permanent process.

Add only that this epistemology-metaphysics needs to be broadened from its overly anthropomorphic focus, not only in Barthes but in most modern philosophy. For all other things also have the forms, the in-form-ations, by which all engage in these mutual transformations that constitute the evolution, that is to say, the continuous creation of the universe. And it is by engaging in this process that all know both themselves and all others equally and daily engaged in cosmogenesis – whether each is in itself conscious of this, or not.[13]

It is in this kind of philosophical context – and not in the context of that rather dessicated territory of Humean sceptical empiricism – that linguistic analysis provides access to the means and sum of our knowledge of reality. Since it is by language that we communally (that is to say, in communication) think how we do, and engage in the practical project that Barthes names as, *agir les choses*. And it is necessary to remember once again that other species use languages; nor need we bother to add, sign languages, since any system of signs is equivalently a language. That having been said, there is little if any conceptual difference between language designed to do things and the idea of word as the operative process of continuing creation. Provided once more that word is taken, not in the sense of expletive-like commands, but rather in the sense of intelligible forms or formulations inscribed in various ways in, and expressed in various ways by all things that make up the creation.

Put in these terms, the question that lies in wait for all truly critical, unprejudiced analysis of the languages by which we investigate all engaged in the cosmogenic process named *agir les choses* is this: what number or kind of such cosmogenic agents do we encounter through the traces they leave as they continually

---

13. Remember Southwood's question referred to in note 11 above: how do they know?

co-create this universe? As we pursue this unavoidable question, this quest, we do of course notice how the very language games by which we both discover and co-create this evolving universe, in both the primal imaginative and the more derivative analytic forms, betray us out of their very own strengths and lead us astray. Imagination in its attempts to construe the world, in both senses of that word, construe – to represent and to co-create – can easily overstep its limits and imagine unreal and unrealisable states of being – the imagination carrying us towards the merely imaginary.

Analytic, that is to say, abstract thought, on the other hand, can take its abstractions and some of their logical developments for whole and concrete realities. Then, in forgetfulness of the fact that these have been abstracted from the concrete, ever-changing flux of being and non-being that characterises the universe (Heidegger's forgetfulness of being), these can easily come to be thought of as things in themselves, forms immutable and universal, and sometimes as existing outwith the universe as really real, ideal and regulative reality. The Pre-Socratics pointed to the characteristic failures and excesses of imagination in ancient myth. And then began their preferred development of analytic, logical reasoning, known as philosophy, in order to ameliorate matters. Postmodernists and some of their predecessors pointed the accusing finger at the characteristic failures of philosophy, particularly at its metaphysical depths. And since many of these failures involved the reification of abstract forms or ideas in order to appear to arrive at divine-like entities squatting outside the world, they denounced them as onto-theology. But there are many other examples of the reifying of the abstract fruits of analysis and its accompanying logic, such as those found in 'Cartesian' and equivalent kinds of dualism. And although some of these do not seem to be examples that would fit the category of the divine, they need to be denounced also, this time as onto-philosophy. Even and perhaps particularly when they occur in the works of the postmodernists themselves.[14]

---

14. Fuller details on this critique of onto-philosophy/theology is provided in my *The Critique of Theological Reason*, especially ch 2. An earlier criticism of an instance of onto-philosophy occurs in Marx's critique of Feuerbach. Accepting what he thought to be Feuerbach's reductionist move, that of retrieving the attributes of divinity for humanity, Marx then complained that Feuerbach's humanity thus defined, still seemed

CONCLUSION

The real question behind and within the subject matter of this piece concerning creation and transcendence, is not: is there (a) god? But, rather, what or who, if anyone or anything, amongst all the entities that present themselves through the traces they continue to leave all around you and indeed in you, do you think could reasonably be called divine? The answer from the Christian theology of creation – and the extent to which that is primarily what is called a natural theology, that is to say, one accessible to natural imagination and reason has strong biblical backing[15] – the answer is: some forming reality that forms like minds form words, but that is not itself formed, that is to say, shaped, cut out, limited as are all the formed entities in and through which it continues to act. An ur-former that continues to produce the endless play of these forms amongst themselves, which in turn produces the ever evolving universe, orchestrating the playing of these so that the evolving universe continues to be co-created by them, and so always and everywhere accessible to all through such cosmic traces.

At the beginning of his quest for *Being and Nothingness* Sartre accessed an entity in some respects similar to the Christian creator: a consciousness-type being in itself contentless, that is, without form or forms which he calls an absolute, the *pour-soi*. But then he did, as already remarked, deny that this or any other such consciousness-type being could ever function as creator of the world, since nothing to which it could give rise – if it ever could give rise to anything – could ever escape from its own subjectivity and so be described as in any manner or form an objective reality. There are many things that could be said about this posi-

---

to squat outside the world rather than seem an immanent co-operative part of it. However, what Marx failed to see through the smoke and flames of this latest blast at an implicit onto-philosophy was this: the attributes which Feuerbach retrieved were such as identified divinity – from his opening attribution of transcendence-infinity to human consciousness – and would become no less so in Marx's corrective resort to full immanence, since the world to Marx is humanity's extended body, and humanity, the 'species-being,' is the world's continuous creator.

15. However one translates John 1:9 in the context of that famous prologue, it is unavoidably obvious that we are being told that the Word which became enfleshed in Jesus of Nazareth is the divine Word that creates the world and that (thereby) enlightens everyone at all times and places in the creation, presumably concerning itself and the world.

tion, and some have been said already. But just one more thing needs to be said about it in order to bring this long analysis and argument to some conclusion. Subjectivity and objectivity are here assumed by Sartre, as by so many others who bandy these terms about, as dichotomously distinct and opposites, whereas they are in fact co-relates or co-ordinates, like immanence and transcendence. All knowledge is subjective in that it is essentially a function of a subject. And it is simultaneously objective in so far as that which is known, and in so far as it is truly known, exists independently of any current instances of awareness of it.

Or, as Hegel pointed out at the beginning of his *Phenomenology of Spirit*, the distinction between subjective and objective, denoted in the metaphor of inward and outward and cousin to the duality of immanent and transcendent, falls always within the subject. This is most clearly illustrated in the case of our consciousness of other selves, which can only take place in the context of some shared inter-subjectivity.[16] The objectivity of these other subjects, of these other subjectivities, is not only the most assured instance of our sense of objectivity, in that their freedom and independence assures us that they are not merely parts or extensions of our selves; it is also the ground and principal guarantor of the objectivity of other things that we jointly know. Add now the consideration, already canvassed, that it is much more feasible to imagine mind-like beings giving rise to matter, in that it is our scientific experience that forming (shaping and thus limiting) gives rise to space-time and thus to what might be called the materiality of things, rather than try to imagine on a cruder materialist conception of the world how mind-consciousness could derive from this, and the answer to Sartre is complete.

So, if the traditional Christian ideas about divine creation were to be put in the manner already outlined, and if we were to opt for the epistemology-metaphysics of, say, Barthes rather than, say, Derrida – traces continually left everywhere by agents forever active and acted upon in the process of cosmogenesis,

---

16. This perception is an intrinsic part of the other perception already referred to, namely, that signs exist only in the context of consciousnesses-in-communion. The quest for the source of this communion of individual consciousnesses is close to the start of Indian philosophy and its road to the Absolute. See Ananda Wood, 'Objective Pictures and Impersonal Knowledge,' *Studies in World Christianity* IV. 2 (1998), 187-211.

and continually read by those agents who are thereby conscious of all engaged in the task defined as *agir les choses* – then it would be difficult indeed to understand why religious folk and some postmodernists[17] are prevented from making common cause in the quest for the ultimate source and conductor of the grand enterprise of being.

---

17. For this is a cause common also to an increasing number of contemporary scientific cosmologists who are working away at their task of discovering the deepest reaches of the cosmos, in general without paying a blind bit of notice to postmodernism, at least in its extreme Derridaen form. For it seems that we are indeed witnessing a relatively new phenomenon in recent times: in defiance of persistent missionaries of allegedly science-based atheism such as Richard Dawkins, but with as little trust in the god-of-the-gaps strategies that some scientists and philosophers, but mostly theologians use in defence of recognisably religious views – for such strategies depend upon the uncertain and unsatisfactory coincidence of finding features of the universe that science does not now explain, with the hope against hope that science never will explain these – an increasing number of scientists express the conviction that when they, in the name of science itself, lift their discussion of the traces of agency from the more specific levels (e.g. the origin of life or of consciousness) to the more generic or universal, that is to say, to the 'meta'-levels which follow in unbroken heuristic sequence (referred to above as the epistemological/metaphysical levels), then the agency engaged in the project, *agir les choses*, may well begin to be perceived, however dimly, to be deserving of religious aura and awe. So that, as my erstwhile colleague at Edinburgh, the physicist Wilson Poon put it, 'the commitment to truth which leads a scientist like myself to the laboratory is, ultimately, not different from that commitment which issues forth in the reasonable activity known as worship.' See Poon's severe but just critique of Keith Ward's god-of-the-gaps approach in his review of the latter's *God, Chance and Necessity*, together with some notice of writers who adopt a more integrated approach, in *Studies in World Christianity* IV.2 (1998), pp 255-258.

PART THREE

The Role of Imagination in Science
and in the Theology of Creation

CHAPTER FIVE

The Role of Imagination, the Instance of the Fictive, in Straight Science, and in Theology*

For my contribution to this conversation I have chosen to reflect upon a role for imagination, and hence a place for the fictive, in all our knowing, in all knowledge. And it seems most apt to do this for this occasion by concentrating on the concept of creation. For this is a concept that has long engaged both scientists and theologians, though not always with positive and harmonious results. For the fictive, the imaginative can also be called, and correctly so, the creative element in knowledge, or the creative kind of knowledge. Certainly the fictive is the creative element in the expression of knowledge, in the creative kinds of the formulation and communication of knowledge. Correspondingly if, as I must maintain, creation as a cosmological concept actually takes the form of that evolutionary process by which our universe continually comes to be, ever reforming, deforming and transforming itself, then undoubtedly creation as a cosmological concept would itself suggest that what we call the creation can be properly known and expressed only by those who participate in its continuously creative evolution – and therefore only by a kind of knowledge and expression which itself has an in-built creative element. I leave to other participants in this symposium, who have devoted more of their professional time and intellectual energy to the study of science fiction, to address the issue of the relationship of those who write science fiction to those who publish 'straight' science. But I should imagine that each degree of success in securing the fictive, creative element in all knowledge and then, naturally, in 'straight science' also, would reveal a similar degree of affinity between the work of straight scientists and of science fiction writers – an affinity that might well be extended to the work of theologians also; thus to facilitate a more constant conversation between all three groups.

* First presented as part of a Conversation between invited scientists, science fiction writers, philosophers and theologians, held under the auspices of the John Templeton Foundation in London, 24-26 June 2000.

As dialogue partner in this simultaneously cosmological and epistemological theme I have chosen, on the side of science, David Deutsch and in particular his recent book, *The Fabric of Reality*.[1] I choose Deutsch mainly because of his explicit attention to a factor which he calls knowledge, as one of the foundational factors or processes which make up (in the sense of continually re-making) the reality of the universe(s). Indeed, of the other three factors in the fabric of this evolving, continuously creative reality – quantum particles, evolution and computation – the two latter are creative processes primarily through their knowledge-bearing, knowledge-transmitting and knowledge-increasing properties. And that, I think, is the best illustration from the 'straight' scientific side of the correspondence of the creative evolutionary process with the creative knowledge transmission process, to which oblique reference has already been made above.

Within this choice I concentrate further on Omega Point Theory, as it is sometimes called. This choice of concentration on Deutsch's account of omega point was persistently queried during the actual conversation of which this present version of my contribution to the conversation is a somewhat revised result. But I retain this further concentration of focus nevertheless, for the following reasons. I am well aware of the fact that Deutsch's omega point theory is based on the assumption that the universe which originated in a Big Bang will then end in a Big Crunch, yielding the satisfactory symmetry of a big bang in reverse, as it were. And I am also aware that many of the other physicists and molecular biologists who also extrapolate from current accounts of the universe towards omega point scenarios, favour an (almost?) interminable end scene over a big bang. An end scene in which an 'open' universe will become a cold and dark universe of ever more thinly spread black holes and the burnt out cinders of stars. A universe that ends, if it ever ends, 'not with a bang but a whimper,' as the poet put it. But some of these scientists who paint this alternative picture of the end of the world have, like Deutsch, a role for intelligent life which may be in some ways the evolutionary successor of the intelligent life of *homo sapiens*. And some of them can even describe the role of that intelligent life at omega point as omnipresent, omniscient and omnipotent

---

1. David Deutsch, *The Fabric of Reality*, London: Allen Lane, The Penguin Press, 1997.

– adjectives which can be fitted to Deutsch' s account of our successors' end-time enterprises also.[2] So that these scientists also
corroborate, incidentally, the ubiquitous and central role of
knowledge/intelligence in creation past, present and future.

But since my main interest consists in showing how these
omega point scenarios, like origin scenarios, tell us something
about the foundational, and therefore ever present, processes of
our continually creative and created universe – a point I will
make also about the Christian theologies based on Christian
creation stories, both protologies and eschatologies – I am not
really interested in the details of the kind of existence we shall
enjoy in the universe(s) which our successors(?) may (co?) create.[3]
I prefer in this respect to maintain my theological tradition's advice that in such matters of detail 'eye hath not seen nor ear
heard'. And indeed, the more sensible of the scientists who engage in this kind of extrapolation are often anxious to observe
that the universe they study, and in particular its 'dark matter',
may yet have surprises in store for all of us.

As further dialogue partners for these fundamental moves
and entailments made and incurred by Deutsch – as for the
other scientists whose omega point scenarios Mary Midgley
studied[4] – I use a Christian theology of biblical stories and scenarios which takes creation to be a concept or image used from
the beginning of the Bible, for origin and eschaton, and for all
development and reversal of backsliding (redemption, healing,
salvation, and so on) in between. The point being that, as in the
case of the omega point scenarios just mentioned above, these

---

2. For an account and somewhat acerbic critique of these other scientific
'prophets,' see Mary Midgley, 'Fancies about Human Immortality,' *The
Month*, Nov 1990, pp 458-466.
3. It is to the kind of detail indulged in by the end-time scenarios she ex-
amines that Mary Midgley's acerbic critique is mostly directed, as well
as to the view that it is in the role of eternal agents of science and technology that our eternal destiny consists; and to the conceit that a species
which is making such a devastating ecological mess in our world
should allocate to itself alone the task of one day re-designing the
whole universe.
4. These included Freeman Dyson who, on the occasion of this conver-
sion, tried to distance himself from, if not to disown his own published
omega point scenario of a decade or so before. But since he had done
nothing like this in the years between, his self-distancing was not entirely allowed on this occasion.

protologies and eschatologies really tell us what we can know about the structures and processes of a continuously created and creative universe at every present moment from the first (if there was a first moment) to the last (if there will be a last moment).

So then, to these matters in the order indicated.

I. IMAGINATION, FICTION AND KNOWING

The role of imagination in the process of knowing is best supported in general epistemological terms by the more sophisticated forms of pragmatism, or what in British philosophical territories and their cultural colonial dependencies is termed the practicalist theory of knowledge. This is a theory to the effect that 'knowing that' is intimately related to 'knowing how'. In such a manner that knowing how to do something, or to have to do with something, maintains a general priority over knowing that something is, and is about to be, really and truly the case. In common language, we learn all we will ever know about reality by creatively (or, as might just as easily happen, destructively) interacting with it.

That is to say, we live in a universe, not of things in some neutral, factual, purely objective sense of that word, but rather, as Heidegger's choice of the Greek word *pragmata* for things was intended to imply: we live in a universe of things with which inevitably we have to do, as each of these things equally inevitably has to do with the others and with us. In this inevitable and universal interaction all of the co-involved *pragmata* are reformed, deformed it may be, but regularly transformed in any case. And in that very process they mutually *inform* each other. In-form being a term that corroborates the view that it is in knowing how to have to do with the things of which the universe is continuously made up that each gains its essential access to knowing that the universe and the things that continually act to make it up are as they presently are, were as they were (by backwards extrapolation) and will be as they will be (by forwards extrapolation) – irrespective of whether the resulting knowledge occurs at the pre-conscious, conscious or intelligible levels. And since this foundational knowledge of reality is gained in and through the processes of mutually and inevitably having to do with each other – gained, that is to say, through the actual transformations which this mutual interaction always involves – then this foundational knowledge must contain a creative, envisaging, imaginative, fictive element. As each interactive agent trying to maintain and if possible enhance its existence in these inter-changes, inevitably tries to know how all of this interactive exist-ence can be for the better for all.

This general epistemology is most fully secured, as it hap-pens, by the most thoroughgoing evolutionary account of the whole fabric of reality and of its whole history. The kind of evol-utionary account I have in mind is that favoured by Deutsch and, I gather, an increasing number of scientists. First, in de-tailed accounts of evolutionary processes, adaptation is placed prior to accident as the evolutionary engine that in the course of understanding and explanation takes prior place to accidental genetic or other mutation, now seen more as occasion rather than cause. And, second, this concept and theory of evolution is extended beyond the more narrowly defined domain of life and living things. For life itself evolved from the mutual adaptation to each other and to the environments they created for each other, of the formed entities that existed before, or at a 'lower' level than that at which life began.[5] Deutsch rightly describes this process of creative adaptation, called evolution, in which the whole continuous creation of the cosmos, from alpha to omega points consists, as a knowledge-bearing, knowledge-de-veloping process. And in this way he corroborates the practical-ist theory of knowledge, at the same time as he (at least implicitly) confirms the central role of imagination or the fictive, the cre-ative, within it.

For every active entity in the entire, co-ordinated evolution of the only universe we directly experience, knows the other en-tities that are integral, with it, to this universal process, by their creative interaction, that is, by their mutual active adaptation to each other and to the environments they create for each other. And, furthermore, this 'knowing' entails an essential act of 'en-visaging', a constant condition for the creation of the new or modified forms that are always necessary in order to envelope in the adaptive process the physical or genetic mutations, or other new 'arrivals', that are experienced in the immediate eco-logical niche of the creative-imaginative, adaptive entity in question. *Homo sapiens*, the most advanced knowledge-bearer

5. See Aristotle's definition of nature, *physis*, as form which has within itself the source of *kinesis*, 'movement' or change, mutation – whether *physis* refers to the whole, or to the smallest part of the cosmos.

and knowledge-developer amongst those that we currently apprehend, can equally be said to know as much as she knows about the whole continually creative cosmos, courtesy of that self-same process by which we learn about the imaginative adaptations of other entities. One thinks of modern medicine, for example, learning to its consternation how certain viruses have developed immunity to our antibiotics, and others, like the ebola virus, steal a march on us so successfully that we may have to learn to live with them rather than hope to destroy them. All courtesy of the self-same process by which we then creatively envisage in the course of these encounters, how we can live better and better, while preserving and enhancing, rather than diminishing, the cosmic environment in and from which we are destined to live for as long as we may.

The unity, and the consequent universal significance, of these knowledge bearing and developing agencies across the obvious differences between physical, biological and intelligent sectors, is secured for Deutsch by his deployment of the common metaphor of computing. Cells are computers running the computer programmes inscribed in genes, and this interactive process in which all engage is 'more than just computing. It is virtual-reality rendering'.[6] Our brains are computers, though currently computers of the more classical kind, whereas the quantum computers of the (near?) future will have their programmes inscribed, not on silicone or current kinds of brain tissue, but on the particles which, together with the knowledge transmission and knowledge development that is the point of evolution and the effect of computing, make up – in both senses of that term: to constitute and to constantly create – the whole and entire fabric of reality. This image of computation as the unifying metaphor for all knowledge bearing and developing agencies that thus engage in cosmogenesis, instead of merely engaging in some more localised and fragmentary enterprises that could never yield a cosmos, is most clearly seen in Deutsch' s account of omega point, although I shall argue that this must then constitute an account of all that is already going on in the past and continuous creation of the cosmos we currently know.

---

6. Deutsch, *The Fabric of Reality*, p 178. Virtual reality rendering in this context must refer to the creative element involved in the mutual adaptation in which the evolution of a continuously creative and created universe consists.

I must also observe that behind this unifying metaphor of cosmic computation, which seems to secure such unity for the cosmic factor or process Deutsch names as 'knowledge', in the making-up of this evolving, continuously created and creative universe, lies the question of unified universal agency responsible for an evolving universe. Cells are computers, human minds are computers, according to Deutsch, but universes do not come into existence or remain universes courtesy of the mere contiguity of myriads of such individual computers. And certainly not by having each of these running its own little programme for its own adaptive welfare. Perhaps it is to Deutsch's fictive account of his omega point scenario that we must then look, in order to see what kind of agency is envisaged as the compute-er that has the ability to create a true universe, then as now, and also up to the present moment. Keeping in mind all that has been said about extrapolated protologies and eschatologies that are actually designed to tell us as much about what goes on in every passing now, as in every past and future.

II. SCIENTIFIC IMAGINATION AT OMEGA POINT

Deutsch accepts Tomasso Toffoli's remark that in the course of all this computation, 'we just hitch a ride on the great Computation that is going on already;' while all the while insisting most strenuously that we do not then see ourselves, as some sociobiologists and evolutionary psychologists see us, just as 'someone else's program running on someone else's computer.'[7] Deutsch then imagines omega point by imaginatively extrapolating from the current cosmic concourse of adapting-envisioning agencies as described in his 'theory of everything'. This theory, it will be remembered, rests on the observation-conviction that quantum particles, knowledge, evolution and computation – these four and only these, in concert – are the equally real and foundational factors or processes which make up the entire fabric of continuously creative, evolving reality.

Then, solving to his own satisfaction the problem of providing sufficient energy to run a universal virtual reality generator,[8] and assuming a development of sufficient power of intelligence (amongst our successors?) to accomplish the task, Deutsch goes on to describe the creator of a universe that offers the prospect of

7. Deutsch, *The Fabric of Reality*, p. 346.
8. See Chapter III, p? above

eternal life. 'With one second, or one microsecond, to go (before the Big Crunch), they (our hugely more intelligent and happy successors) will still have all the time in the world to do more, experience more, create more – infinitely more – than anyone in the multiverse will ever have done before them.'[9] The creation of this end-time eternal universe (eternity coincident with a microsecond of current cosmic time) Deutsch – in common with those whose omega point scenarios envisage not a bang but a whimper – conceives to be the work of a co-operative community of distant successors of *homo sapiens*. And precisely because these creators are conceived as being many, not one, Deutsch can insist that his fictive account, following on his extrapolation from known cosmology, gives no support to the traditional Christian idea of one God, Father almighty, creator of heaven and earth – even if it be intended as a form of support only for the emergence of such a god at the end rather than the beginning.

Yet it is worth observing even at this point, and before offering a very brief sketch of a more critical Christian theology of the creator, that these end-time creators of Deutsch's eternal universe(s) must act as one mind if they are to succeed, their mutual criticism as they envisage and experiment acting purely as the on-going self-critical function of any mind engaging in such creativity. For the rest, as I must shortly say, there is no incompatibility in the Christian account of continual creation, between the foundational, unifying, restorative and evolutionary efficacy of the one creator God, and the similarly describable efficacy (but derivative rather than foundational) of the co-creative activities of the members of the human race, as indeed of all the other co-creative *pragmata* which literally make up the evolving universe. And Deutsch's (and others') end-time creators, however close their creative power may come to full participation in that of the traditional creator God, are still in their existence and efficacy derivative rather than foundational. For they are derived from this current universe and, furthermore, their 'new creation' (a term used also in Christian scriptures) will be made out of the shards of a shattered (or slowly 'dying') universe, thus derivative of the enduring forms and properties of the matter still available from the current universe.

---

9. Deutsch, *The Fabric of Reality*, p 352.

III. THEOLOGICAL IMAGINATION

I have argued at length elsewhere[10] that the great Judaeo-Christian story of creation, the creation myth which seems to occupy but the early pages of Genesis, but which in actual fact is rehearsed and reworked from beginning to end of the Bible, is not centrally concerned with the remote origins in the past of everything that comes into existence. Interest in such a remote act of origin, its 'time', character and conditions, is sometimes a minor concern-in-passing in this as in all great myths of creation, but that is all. But nor would it be adequate to say instead that this myth is centrally concerned rather with the way all things are as they continually come to be in the ever-fleeting present. For the mighty myths attempt to envisage how things are, from the point of view of their ever-active source, from the point of view of their *fons et origo*, in that sense of the word 'origin'.

Furthermore, these myths seem to feel that they must, logically, attribute some kind of fullness, in the sense of perfection of being and living and knowing to such an awesome and permanent *fons et origo*. So that these cosmic myths must then attempt to see simultaneously how things are and how they ought to be, if they are, as it were, to live up to their eternal sourcing, to the pristine perfection that such a source ever intends and offers. It follows then that all the great myths of creation are in fact morality tales. At least in this, that in envisaging the ultimate creative source, they inevitably also try to envisage the on-going creation as it must be intended by, and look to the benevolent source of all this existence and life. So even in the early Genesis story, there is an evocation of the idea of entities with which we have to do co-operating in the creation of our world. Humanity in particular, created in the very image of the divine creator, are appointed stewards of other things which bear the seeds of their future, each after its own kind. And correspondingly there is the element, in the knowing of all of these entities, of envisaging, imagining how they are and how they ought to be together. Not just of 'scientific' noticing of how they happen to be at any particular instant. The fictive, again and always, is an essential feature of the knowing of reality, as it is and as it ought yet to be.

But this in turn does not mean that the central concern of

10. See my 'The Creator, the Scientist, and the End of the World,' in *Studies in World Christianity* rather than the much revised version of this in Chapter III above

such myths is with some instance of perfect, 'new' originating that belongs to a remote point of time in the future, any more than it is with the 'new' originating that took place at a remote point of origin in the past. Even in those apocalyptic forms of eschatology such as we find instanced in Deutsch, those extrapolations from current knowledge that seem to focus the central concern of creation myths on the remote future, concern with a remote end in the future is as minor as is concern with a remote point of origin in the past in the creation stories that concentrate on the rich metaphor of beginnings. The effort, rather, is to see in equally fleeting past, present and immediate future how all things in a constantly created universe relate to each other and to any suspected source of all. In such a way that all of this mutual adaptation, in which the constant creation or evolution of the whole fabric of reality consists, is to the benefit and not the detriment of each envisaging-active agent. Although there is of necessity, as it would appear, an element of being and living more fully at the expense of the other, in each and every mutual, knowledge-bearing and knowledge-developing adaptation, the aim of creation stories is to envisage simultaneously how things are and might (or ought) to be to the benefit of all, even if this does entail for all an ethic of self-giving, as a Christian might put it.

Finally, stories of creation, like the Genesis stories, do sometimes tell of envisaging-adaptive agencies acting occasionally, not for the mutual benefit of all, but for an advantage of some which will be excessively destructive of others. And as these fall stories then inevitably call for, or point to healings (salving, salvation) and restorations as essential parts of the continuous creative process, the same must be said of these parts of the extended creation stories: destructive falls from goodness described as past (in the beginning), and full salvings and restorations apparently postponed to the very end of the future, are also in fact descriptions of what happens, or may do, in every fleeting present.

One further point needs to be noted about the Judaeo-Christian creation story that is being formulated and re-formulated through the whole of the Bible. It is this. The knowledge/intelligence which continually creates this universe, and which in this story is naturally pictured as a personal intelligence, does so by acting through the other agencies that its formative creation-processes successively cause to evolve, and most particularly through *homo sapiens*. This is made clear especially in the

Christian doctrine of the incarnation of God. The Word by which the world is continually formed, and that thereby enlightens everyone in the world, is incarnate in a human being like us. And this insight is expanded, for instance, in the Pauline literature of the New Testament, as Christians call it, when the whole of creation is seen to be in bondage, awaiting the creative liberation of the sons of God. For it is we humans, the ones the Bible calls sons of God like Adam, who keep so much of God's good world in a barren and ruinous state that could very conceivably prove the ruin of all life on this planet. In the words of the poet, Brendan Kennelly, so powerfully describing 'man's power to corrupt':

Out of his cage he sends his cries,
Impressing his slavery on free things.
Fields, cities, seas
Are stained with his yearning.
His blood stains the summer sky
And he wonders why
He tries to glorify
Himself
By turning the rattle of his chains
Into the music of a word called freedom.

(Who, on reciting these verses would not think of Bush and Blair busy exporting 'freedom' to the unfortunate Iraqis, and the death and destruction and daily terror and misery their immoral war-mongering brought to that tragic land, and the rape of the land itself, and the blasphemous suggestions by both of these men, that their faith in God was at least part inspiration for these foul deeds?)

So Paul envisages liberation from this age-old enslavement of the world to evil, selfish, destructive ways, by those who have begun to live by the spirit of the creative Word that Christians say is incarnate in Jesus; but that is also always working in the whole course of creation to enlighten and inspire to creative goodness all who are born into this world, and who through this inspiration would aim at the stature of re-creative sonship.

The ultimate creative source, in short, acts utterly immanently in and through all the continuously created and creative agencies that make up the unity of this universe, and acts in and through humanity *par excellence*. This intelligence is of cosmic efficacy and extent. For a cosmos, a universe, could scarcely be thought to emerge at the beginning, or now, or in the future, from the adaptive enterprises of a pure plethora of more localised knowledge bearing and developing agencies. So that the existence and creative activity of the cosmic intelligent spirit is also commonly said to be utterly transcendent. But this can only mean that it transcends from within, or that it acts from a position of immanence, through and beyond every stage of the continuous creation or evolution at which each and every other derivative agency successively finds itself. The creators of eternal universes that Deutsch, together with some proponents of an eternal-whimper-universe, envisage as a race of future humans, appear as creators of universes in the same sense and to the same extent as the creator presented in the Christian myth of creation. Except that, in the Christian myth and that of many other religions, humans appear to be what in fact they are, namely, co-creators with the ur-creator, if only because of the logical need to account for the creation of the universe during the aeons before *homo sapiens* evolved as an intrinsic part of its evolution.

<h3 style="text-align:center">IV. THE POSSIBILITY OF DIALOGUE</h3>

What, then, does the (Christian) theologian have to say to the scientist concerning their respective uses of imagination in their stories of creation? (Apart, that is, from the obvious admission to the effect that the Christian story of creation inherited from Biblical times, is cast in the out-dated cosmology of that era.)

First, that Deutsch's account of the activity of our end-time intelligent evolutionary successors is truly a creation myth, a story of the creation, from the particle-shards of this one, of a universe of which eternal life is a feature. By myth, it may be no harm to repeat, in view of the word's pejorative usage in modern discourse, is not meant a fantasy that is simply infantile in terms of the life-history of the race, something fundamentally false. Quite to the contrary; myth is a product of the imagination functioning, as it does, in a process of knowing which essentially entails envisaging the best outcomes of adaptive interaction. We, members of the human race, are what we constantly become, in a universe quite literally made up of agents, of all of which that same can be said. Hence cosmic myths, which usually acquire that title because of the depth and comprehensiveness of their accounts of our world, can be true or false or, more often, a mixture of truth and, if not falsehood, at least limitations or distortions of the truth, just like any other accounts of reality, scientific accounts, abstractly analytic or philosophical accounts, and those inextricable mixtures of all of these that are found alike in common, scientific and philosophical discourse.[11]

Second, that to call this universe that is created in the apocalyptic end-time a virtual-reality universe, is to mean no more than was meant when Deutsch said that cells engaged in computing are thereby engaged in virtual-reality rendering. That is to say, the term 'virtual' now refers to the knowledge which Deutsch regards as a constituent element of the fabric of reality, and it refers in particular to the 'know-how', the imaginative-fictive-creative feature of that knowledge, that intelligence, that knowledge bearing and developing agency, at the point when it reaches truly cosmic dimensions of creativity, or, when it is inscribed upon, actively immanent within, the most fundamental structures of matter. By contrast, what we now commonly, if more restrictively, call virtual reality refers to creative or fictive knowledge or experience in the form of computer programmes inscribed upon grosser forms of matter, such as silicon, and thus producing merely artificial environments.

Third, that on this obvious understanding of the term 'virtual' in this context, and on Deutsch's own logic, a fictive-knowledge, a creative intelligence at least equal in (infinite) capability to that of our imagined successors of the Big Crunch era, must be the source of the (virtual) reality of this current universe of our direct experience. And it must then be responsible for the creation-evolution of us also, as indeed for the co-creative contributions we make now or in the future. For it is, after all, as an inherent part of the intelligence-driven continuous creation or evolution of this universe that we lately emerged and now evolve to whatever degree we do evolve our characteristic powers of co-creativity. And, further, it is from the already formed matter of this universe, however violently deformed by the convulsions of the Big Crunch, or otherwise thinly spread and depleted in the Long Whimper, that our successors will (co)create Deutsch's brave new world, or whatever other more wimpish world their new-found omnipresence, omniscience and omnipotence may secure.

---

11. I have written a little more about the nature of myth, as myths occur in both the scientific and religious arenas, in 'Christianity: Critiques and Challenges,' in Philip Esler (ed), *Christianity for the Twenty-First Century*, pp 25-42.

This illustrates in the case of Deutsch and other omega point theorists also what is meant by saying that the end-of-time (beginning-of-time) scenarios really tell us something about the foundational factors and processes that operate in every fleeting present.[12] In Deutsch's terms knowledge, as one of the four factors that actively make up the evolving fabric of reality, acts as a creative kind of computation, both borne by and resulting in the very stages of evolution, because its constantly changing programmes are inscribed within the material element (in quantum theory, in the particles). This Great Computation, as Toffoli called it, eventually brought *homo sapiens* into being, and *homo sapiens*, far from being just someone else's programme running on someone else's computer, increasingly participates in this cosmic computation, until eventually *homo sapiens* is, in and through this co-creative activity, at one with the Great Programmer – in some yet-to-be-determined sense of 'being at one with'. That is to say, the details of this final state of the uniting of *homo sapiens*, and perhaps of all other evolutionary agents, each according to its kind, with the agent of the Great Computation, as well as the details of the resulting universe(s), may escape us. And they tend to change in any case in our efforts to imagine them in terms of our changing comprehension of the cosmic processes we are and have always been talking about. See, for instance, the difference between the accounts at the opening of Genesis, of *homo sapiens* as steward of creation or, in the second creation story in the same text of Genesis, as the gardener in the garden, and the accounts of the role of the presumed evolutionary successors of *homo sapiens* in the various contemporary omega point scenarios. But in all these differences, whether they occur in end-time or beginning-time scenarios, what is claimed to be revealed, and to be described with varying and, we hope, ever increasing degrees of success, is the complex of creative factors and processes that bring universes into being in every fleeting present, and in any form of time or eternity.

So what are these factors and processes, differently glimpsed and described at different times? In the attempts to answer that question lies the most fruitful prospects for dialogue between

---

12. Of relevance here is a remark which Stephen Clark made during the symposium discussions, when he suggested that Stapleton was not really writing about a long time ago, or a long time hence, but about 'now'.

scientist and theologian. The theologian – at least the one steeped in the tradition of Greek-inspired Christian theology of creation – will point out that the truly creative agency named simply as 'knowledge', active in the forms and processes of computation and evolution, raises metaphysical questions about its status as such a cosmogenic agent. It will not do simply to keep on calling it 'knowledge', without having to say further what kind of entity it is that does this kind of creative reality-rendering. Clues are given, however inadvertently, by omega point theorists when they talk of powerful intelligences acting as one mind to create eternal worlds. But the story of our universe, in which *homo sapiens* appeared relatively lately prevents our taking these clues to point simply to our race in however evolved a future form. A unified source intelligence of more than cosmic size and might, metaphysically prior to our race and kind, is indicated already, no matter how much our present or future members may come to participate in that cosmic creator's activities. And once *homo sapiens*, however hugely involved, is seen as participator in, or agent of this metaphysically prior knower, the prospects of success in creating and inhabiting the best of all possible worlds may be considerably advanced.

There, in any case, is the source and subject of a necessary dialogue, one which has in fact gone on in Western philosophy from the beginning, when what we now call science, philosophy and theology can be seen to be simply names for different stretches of one and the same extended inquiry into what the Greeks called the *physis ton onton*, the nature of things.

Fourth, that Deutsch's metaphysics (for that is what his work amounts to) can facilitate much more than more ancient metaphysics some understanding of traditional puzzles: for instance, how eternity can coincide with, or co-inhabit, an instant of current cosmic measurement of time (see his story of the last microsecond of the Big Crunch); and how eternity may therefore be accessible in or to an (any?) instant; or how mind/intelligence can create time (and space?) and so the whole of spacetime which we call the material universe.

Finally – for one could go on, and on – that the theologian can talk to the scientist, and indeed must do so. Not with any aim of dragging theological concessions from scientists unwilling to engage in such, to many of them, discredited practices, but for the simple reason that theology, in trying to imagine the continuously creative empirical reality we share, from source to goal, simply has to work with the best accounts of that reality which the most advanced science can supply. By seeing to it that theology works like that, theologians may at least come to seem less discredited than on many counts they still deserve to be.

CHAPTER SIX

Gerard Manley Hopkins: The Poet as Theologian*

I

Had I two hours rather than less than half-an-hour, I would broadcast three themes in order to prepare properly the ground for the few ideas I would wish to sow in your minds today, in the hope that these ideas might then enjoy the best conditions for bearing some fruit.

First, I would spread the thought that imagination is our prime heuristic faculty, and that the image, and more especially the symbol, and still more especially the metaphor (that drawing of images into those pregnant tensions that give birth to the most depth-sounding symbols), the image is the primary tool of that prime heuristic faculty, the imagination.

An opening excursus of this kind, did time allow, would involve a lot of arid epistemological argument to the effect that our first and last way of coming to know ourselves and our world consists in the simultaneous openness to and construal of those images of reality that carry us beyond the factuality of a momentary snapshot to the potentialities always inherent in mutually interdependent minds and materials. An epistemology of vision, in short, in its suggestive ambiguity of what is seen and what has to be simultaneously envisaged. A theory of knowledge that treats reason, itself imagined as a separate function and faculty, with its characteristically critical processes of analysis and synthesis, as a second-level, if sometimes – often? always? – necessary sifting of the truth value of the visions with which we must begin and would be wise to end.

Some of what I would have to say on this first theme is sometimes in part at least acknowledged in any case, for example, by the one who said – who was it? – that the poets are the unacknowledged legislators of the race. But I would say more, and say it of all our artists. If only because someone else said, equally

---

* First delivered to the Gerard Manley Hopkins Society's XVth International Summer School, Monasterevin, 19-26 July 2002

truly, that the law is an ass, and as I would not want your highest accolade to rest upon your alleged prowess at designing donkeys, I would want to add that the artist, servant and master of imagination, is the first and finest theologian of the race.

Second, and in part explanation of that last claim, I would want to add the thought that imagination and its characteristic vision already encompasses the concept of revelation, and does so at any depth or height of reality we may care or dare to visit. To put the matter in the terms of Hopkins' own metaphysic and epistemology of instress/inscape: some power of being (perhaps what Dylan Thomas called 'the force that through the green fuse drives the flower'?) that gives and sustains in everything its individually distinctive form or inscape (the primordial creative instance of instress), sees that inscape further inform in the imagination of the one open to receiving it a corresponding inscape (a consequent revelatory instance of instress). And then as this perceiver is conscious of the creative-evolutionary co-operation of all of these individual and mutually instressing inscapes, and most particularly of the perceiver's own co-operation with all these others, she construes and expresses, as much as she merely receives, these second in-formations or inscapes – and at her best publishes them primarily as art (a concluding revelatory-creative instance of instress). In so far then as the primordial power-source of all formed entities in formation of a universe is identified as divine – and it is as the constantly operative *energeia*, operative in all forms of life and existence, that the god is always identified – the linked sequence of instress and inscape outlined just now yields the vision of the artist as the primary recipient, custodian and dispenser of the revelation of the divine, the unacknowledged theologian of the race.[1]

Third, and in part development, part defence of that last point about revelation, art and theology, I must issue a caution concerning the word 'Otherness' in the conference title. The term, The Other, is much bandied about in continental philosophy and by its borrowers on these islands, and the presumption seems to be that everybody knows what is being talked about. So much so that, like talk of the emperor's clothes, we badly

---

1. This matter is much, much more adequately developed in the chapter on 'Art and the Role of Revelation,' in my *The Critique of Theological Reason*.

need the voice of the little child to pipe up and, whenever The Other is introduced with capital letters and solemn tones, to ask insistently, 'the other what?' For otherwise the presumption of meaning will be overtaken by the suspicion that mere mystification is afoot, and that analysts of The Other themselves quite literally don't know what they are talking about. Then this mystifying talk about Otherness is seized upon by those who want nothing to do with divinities, as a means of talking about mysterious dimensions of the universe, particularly in its personal dimensions; while it is equally welcome to religious folk who see it, much like The Absolute or The Infinite – also with capital letters and solemn tones – simply as a cover term for God. And finally, such religious talk about Otherness is often linked with talk about utterly supernatural transcendence, so that revelation of or from the divine is pictured as crossing occasionally and miraculously some infinite chasm, thereafter to be handed down as doctrine or dogma, with little or no prospect of people seeing the meaning or truth of it for themselves.

The terms, Otherness and The Other, probably slipped over into philosophy from the cognate discipline of theology, where it is still often embellished and re-phrased as The Wholly Other. It is then glossed by references to infinite qualitative distance between God and creatures. And it is then linked with loose talk about utter transcendence, not the least of this looseness becoming evident in the apparent assumption that transcendence and immanence are contraries, whereas in fact they are corollaries. All of these moves finally conspired towards the overall impression of a god separated from our world by some infinite distance, but crossing that chasm spasmodically and arbitrarily. Crossing the chasm first in order to create the world by a single act of putting the world in existence at the outset, rather than a continual forming of a world, in terms more in line with the linked stages or 'cleaves' of Hopkins' instressing. And thereafter only crossing the chasm to admonish, punish, send down laws, redeem and resurrect, all occasional acts in favour of just one species in creation, us. And all of this in the end resulting in an idea of divine revelation in which one instance of it is natural, in the sense that the very existence of the natural world makes us think that someone sometime must have put it there, whereas all the other instances of crossing over are supernatural, and have to be communicated by some miraculous means in order to fill in the details of the god's intentions for us humans and of the concrete course and condition of our destiny with the god.

These special, miraculous divine revelations are then thought to be conveyed in the first instance to distinctive individuals or groups, thereafter to be spread by word of mouth to other members of the race. Further, different groups identify different discrete instances of claimed divine revelation as being definitive. The Jews identify the Mosaic revelation, the Christians the revelation that came in and with Jesus, Muslims Mohammed, Baha'is the Baha'ullah, and so on. And each one may then try to argue into submission by tongue or pen or sword or in extreme circumstances even to kill off the groups of believers who went before or came after.

But there is another idea of revelation, and of divine revelation at that, one which avoids the mystifying, absolutist images of otherness and transcendence-as-separation. This is an idea of divine revelation that is much more aligned with ideas of creation-as-continual-forming (deforming, reforming, transforming), and therefore much more aligned with the linked levels of Hopkins' instressing as a cosmic-encompassing, indeed cosmogenic process. The source-former, the ur-instressor, itself without particular form – else there would be a limit to its forming, and there appears not to be – fashions each individual thing; and as each in Hopkins' words 'deals out that being indoors each one dwells,' propagating its kind, and all mutually transforming each other in that interdependence by which each finds life and life more abundant at each other's expense, the source of all this pullulating, self-transcending life-forming, as it moves up the levels of inscape/instress to the poet's receptive spirit, results in the revelation of the cosmic form-er, ur-instressor, utterly immanent source of all this continuous transcendence, in and through the co-ordinated in-formations of all these individual things – 'myself it speaks and spells.' So that all together speak and spell the forming source of all. And it is all natural revelation.

The Christian Bible, incidentally, recognises the naturalness and immanent transcendence of this continuous creation-revelation. In the opening hymn of the Fourth Gospel, when speaking of the creator of the world, in the metaphor of the divine Word that took human form in the life, death and destiny of Jesus, then changing the metaphor from word to light, the text describes this fashioner and former of all things at all times as the light that enlightens everyone. The creative power reveals itself in all of creation at all times to all, and the same text further finds that power now so focused in and through the common human nature of this individual man, Jesus, as to make his followers grasp for that elusive image of incarnation, in order to do justice at once to its individual existence and its universal reach.

II

The few ideas I have to plant in this ground so prepared, then, concern the dominant Judaeo-Christian imagery of divine creation, some instances of such imagery in Hopkins' poetry, and some very general considerations of the role and value of poetry in general, and of Hopkins' poetry in particular, in relation to this theological *topos*.

The dominant Judaeo-Christian imagery of divine creation occurs at the beginning of the Bible, the opening verses of the Book of Genesis which, if I were asked (though I cannot think of any good reason why anyone would ask me to do this), I would translate as follows: 'At the outset of God's creating the heavens and the earth, the earth was formless and empty (*tohu we bohu*: the original Hebrew has the very sound of a hollow wind blowing through an empty and desolate seascape), and darkness covered the face of the abyss; and the spirit of God was brooding over the waters. And God said, "Let there be light", and there was light. And God saw that the light was good; and God separated the light from the darkness.' That is to say, simultaneous with the coming of light, the primeval darkness took form. Darkness begins where light ends, and vice-versa. And on that primordial separation, a series of further separations occur, which give the resulting elements and things each its own shape and characteristic content, its own form and inscape. For darkness submerges all form, makes all formed entities disappear, one might say, where light lets them be and makes them manifest. So there is made within the waters, otherwise called the abyss, a separation and a forming, when the firmament of the heavens is raised between the waters now above and those now below. The earth is then formed by being 'separated' from the waters below, giving seas and land, and so on, and so on.

The fundamental image here is that of a spirit, itself unformed or boundless, brooding, hatching the forms of light in a then and always surrounding darkness, birthing the forms of sound, of word and music, in a then and always surrounding silence; sourcing the forms of activity and turmoil in a then and always surrounding peace and rest; creating the forms of solid lands and seas and all their formed furniture in then and always formless and infinitely formable depths and heights – a spatio-temporal continuum as cosmologists would now call it, of an infinitely expandable universe. Furthermore this is a continuous process in which the more particular forms of elements and things engage with the spirit-creator of each and all of them in what we now understand as the evolution of the cosmos – in Genesis terms, each bearing within itself the seeds of its own future – and none more co-creative (or destructive) than the species which has itself evolved to the level of incarnate spirit, in the image of the ever present original creator spirit ever immanent in the world and always actively and infinitely transcending all finite 'cleaves' from within.

No befores or afters then when the focus of attentive wonder is upon the creator spirit and on that other formless infinity which comes into existence simultaneously with creation. Space-time, within which before-and-after, here-and-there apply, occurs in the infinite interval between the two formless infinities, the one positive in forming and promoting existence and life from within, the other negative in marking the moving limits of all that is created, from its position of separation, banishment, without. It is important to note that, although natural opposites, these two infinities because of a common formlessness attract the same adjectives or attributes: still and stillness, in the meaning of both rest-in-peace and of silence; deep and depth, with height sometimes substituted for the positive infinity; dark and darkness even, with this difference that the darkness of the positive source of all is accounted for by the overwhelming intensity of a light that, looked at directly, blinds all mortal eyes, while the darkness of the negative infinity is due to lack of light beyond the limits so far marked by the finite length of light's creative outreach.

It is this imagery, incidentally, which makes whatever sense can be made of that otherwise unintelligible phrase: creation out of nothing. The formless creative-source of all is 'nothing' when the term, thing, as it must, refers to a formed, i.e. limited, finite entity; the creative source is then the no-thing. That is to say, the creator spirit is not an extra thing, added to all the others; it is a

'no-thing,' as the 'way of negation' in the old Platonic philosophical theology that Christians then borrowed, so fully recognised. The negative infinity is also a no-thing; it is the nothingness that hovers on the edges of all the formed, structured, ordered universe. Yet it can still be imagined as the primeval
chaos, darkness, abyss that constantly threatens to encroach
upon all co-operative, evolving entities, in the nexus of disintegration, death and oblivion. So the foundational creation can be
termed creation-out-of-nothing in the dual sense that it comes
from the creator spirit, and that it has the appearance of being
formed out of the residual darkness that comes about simultaneously at the limits of light.

The finite human mind, of course, can gain no proper pur-
chase upon the ur-creator, but it has an analogy or, better, a symbol of this – a symbol being an image which participates in that
which it images forth – in that image of human (co)-creativity already mentioned, for that creativity is due to the specific nature
of human consciousness as self-consciousness.

Briefly, we are conscious of self in the process of being conscious of forms of things and persons other than ourselves; we
are conscious of self in the process of consciously dealing with
these other forms. This consciousness of self being conscious of
others is then a reflexive consciousness, imagined, that is to say,
as bending back over itself and its contents. Thus it shows itself
transcendent of all of these, having reach beyond the limits of
currently lighted forms, and indeed of all conceivable lighted
forms past, present or to come. This at once accounts for its
creativity; it can envisage the currently lighted forms being
multiplied, changed or even absent. And it accounts also for the
sense of currently unlighted reaches of itself, beyond all such
content. Some modem students of the psyche attempt to deal
with that latter extension under the image of the dark unconscious realm of consciousness – Freud for instance or, better still,
Jung with his collective unconscious. Some more ancient
philosophers such as the great Hindu Vedantists, argue that our
experience of consciousness that transcends all actual or possible formed or concrete content, is our dim prescience of the
infinite Self, the true reality deep within all mere appearances
which constitute the appearances of individual selves.

Hopkins deals in this Genesis creation imagery of the Creator
Spirit brooding over the abyss, in his poem *Peace*:

And when Peace here does house
He comes with work to do, he does not come to coo
He comes to brood and sit.

The brooding bird of creation, action out of the stillness.

Yeats does it more fully, again with reference to the human creative origins of great human achievements, great works of art,
and even great destruction, in *The Long-Legged Fly*. Take the
stanza on Helen of Troy:

That the topless towers be burnt
And men recall that face,
Move gently if move you must
In this lonely place.
She thinks, part woman, three parts a child,
That nobody looks; her feet
Practise a tinker shuffle
Picked up on a street
Like a long-legged fly upon a stream.
Her mind moves upon silence.

The genesis of the terrible beauty, out of the silence. (I once read
a line in one of these transient poems that appear occasionally in
the better newspapers: 'music is an arrangement of silence'.)

T. S. Eliot, in *Ash Wednesday*, deals in the same imagery of creation, but now with direct reference to original, divine creation:

If the lost word is lost, if the spent word is spent
If the unheard, unspoken
Word is unspoken, unheard;
Still is the unspoken word, the Word unheard,
The Word without a word, the Word within
The world and for the world;
And the light shone in darkness and
Against the Word the unstilled world still whirled
About the centre of the silent Word.

Hopkins, of course, visits the theme and imagery of divine creation many times. In *God's Grandeur*, in the context of celebrating
the continuous renewal that characterises every morning that
follows the darkness of every night, he directly invokes the
brooding spirit of Genesis:

Because the Holy Ghost over the bent
World broods with warm breast and with ah! bright wings.

In *Pied Beauty*, in praise of God for 'All things counter, original, spare, strange', he says of God: 'He fathers-forth whose beauty is past change.' The imperative of patriarchal politics that must have fathers giving birth need not delay us here; rather the reference to beauty draws our attention to other instances in Hopkins' poetry of the imagery of divine creation.

For if, as the dominant imagery of Genesis confirms, creation consists in continuous forming, and if beauty is a matter of form, then he can say in *The Leaden Echo and the Golden Echo*:

Give beauty back, beauty, beauty, back to God, beauty's
Self and beauty's giver.

And God is then beauty's keeper; where? 'yonder.'

If creation consists in continuous forming and if, in a cognate manner to the case of beauty, goodness is also a matter of form, this time a matter of ever further forming in a line true to each one's inscape-form and all of the interactions of these, transforming, and on occasion reforming where opposite-of-creative, that is, destructive forces operate, then we can interpret as follows the lines which proclaim that the just or good man is

Christ – for Christ plays in ten thousand places,
Lovely in limbs and lovely in eyes not his
To the Father through the features of men's faces.
(*As kingfishers catch fire*)

(For beauty, loveliness, 'keeps warm Men's wits to the things that are: what good means.' *To What Serves Mortal Beauty*)

The brimming over stillness that brings forth the word, the dense darkness that crowds round the bringing forth of the light, the infinite depth that makes to emerge, all true to themselves and to it, good and beautiful things – and that those calling themselves Christian believe was so completely present in a man called Jesus of Nazareth as to risk the concept of incarnation – to it goes back all that is true and good and beautiful in all creation. In the dual sense of going back: is traced back as to its source, and is drawn back to as its goal. Then creation in its curious combination of retrospect and prospect, is revelation. Then also the co-creative forms of the creation tell forth, not only their own interdependent and moving inscapes, but also, at the permanent source and depth of the universal ensemble of these, they tell forth the (to mortal eyes) dark force, the still centre of the whirling world. Margaret Atwood's recent book, *Negotiating with the Dead: A Writer on Writing*, describes the 'essential nugget' at the heart of all creative writing as a going down into the dark depths 'where all stories begin'. For the creative writer, together with other creative artists, is servant to the image-forms that come from the abyss, and master of the formed-images that can guide us back to the abyss – the artist as seer and prophet, the very best types of theologian.

It is worth remarking here, if only because it poses a further interesting question for students of Hopkins, that there is a fuller imagery of continuous divine-creaturely creation, one which shows that there is a far more substantial inter-relationship between the two formless (i.e. infinite) 'no-things', the life-giving and the nihilating, than that so far noticed in this essay, namely, that both being formless, both attract the same attributes: of silence, depth, and darkness. For in fact the source-of-all no-thing is the origin also of the other, negative infinity. The shining of the light brings about the darkness that crowds round its limits or, in more literal terms, the infinite source of all cannot create other than finite entities. But finite means limited in space-time, therefore mortal. And that otherwise nice phrase about being drawn back to the source must then take on a darker tone, for it now entails a journey through that other darkness, death, oblivion. From the worm's eye viewpoint of earthly life, even for those who can see or at least hope in the permanent metaphysical priority of the positively creative infinity, this crucial point of one's forming involves such utter transformation as to utterly destroy the form of human life we now know from the lowest bodily function to the finest sentiments of which we are capable. So that, from this earthly viewpoint, it is difficult indeed to discern whether we are passing through the negative, form-dissolving darkness into the infinite bliss and fulfilment of some kind of union with source-infinity, or sinking without remainder into its nihilistic opposite.

This trial of discernment is known as the dark night of the soul; this rather than some 'arid' spell in the course of some quite distinctive religious experience thought to be confined to those who are then called mystics. (Observe that many of these who describe this experience, who are named mystics by us, never refer to themselves as members of a distinctive group of human beings called mystics. On the contrary, they write of a journey all must make, whether they like it or not, and they write simply so that fewer of us would have to make this common journey in lazy or cowardly ignorance of its context and contours.)

Did Hopkins, then, so deploy the classic imagery of creation as to reveal this further dimension of its on-going reality? Almost certainly so. It is surely there in *Spelt from Sybil's Leaves*:

For earth her being has unbound, her
dapple is at an end, as-
tray or aswarm, all throughther, in throngs; self in self steeped
and plashed – quite
Disremembering, dismembering all now. Heart, you round me
right
With: our evening is over us; our night whelms, whelms, and will end us.
Only the beak-leaved boughs dragonish damask the tool-smooth
bleak light; black,
Ever so black on it.

But it might also be worth the while of those who study Hopkins' sonnets of desolation and other like poems, and more particularly it might be worth the while of those who are inclined to let things like Hopkins' psycho-physical ailments and the various forms of alienation that he experienced in Ireland play a foremost explanatory part in their critical assessment, to reflect upon the fact that ailments and alienations are but our intimations of mortality, the harbingers of death, poignant signs and instruments of all our darkening nights.

III

So much then for the poet, any poet, as potentially at least the unacknowledged theologian of the race. But what of the Christian poet, which Hopkins would certainly claim to be? And what in particular of his hardy line: 'There's none but truth can stead you. Christ is truth'? (*On the Portrait of Two Beautiful Young People*; this line is interestingly misquoted, I think, in the Penguin Gerard Manley Hopkins as 'There's none but Christ can stead you.' p xxv.)

If what is here said about the light that enlightens every man being intensified to the limit of human bearing in Christ, and

about the Word (now incarnate) that ever and still forms and informs in all the world and all its natural and human history,
then that word that finds expression in Jesus' mortal life should
help light the seer's way to envision all that is also revealed always and in everything; a useful, perhaps in some ways necessary aid, in view of the state of nature that is, in Hopkins' words:

> all seared with trade: bleared, smeared with toil;
> And wears man's smudge and shares man's smell.

Though it may also be that, since this distinctive and, Christians
would say, definitive revelation in Jesus of Nazareth has been
filtered through the ages by his self-declared followers, and in
part at least itself seared, smeared and smudged in the process,
the poet, any poet who is a true seer and prophet, might cleanse
of its distinctively Christian smudges the current Christian version of that same revelation that is available to everyman. On
this latter account of the matter, the poet would be the better a
Christian for being a poet, rather than being the better a poet for
being a Christian. But is this the account of the matter by which
Hopkins himself believed and versified? And which would then
substantiate a claim that he is one of our lesser great poets, as W.
H. Gardner called him, rather than one of our great lesser poets?
Many poems and passages would suggest that it is. Some are
ambiguous, for instance, the passage already quoted about
Christ playing through the features of men's faces, and the pas-
sage in the poem, *That Nature is a Heraclitean Fire* and of the
Comfort of the Resurrection:

> In a flash, at a trumpet crash,
> I am all at once what Christ is, since he was what I am, and
> This Jack, joke, poor potsherd, patch, matchwood, immortal
> diamond,
> Is immortal diamond.

Such passages, and others like them, could be read together to
mean that because of the universal creativity of the Word incarnate in Christ, all forms of being are drawn back to Source and
enjoy eternity in some form, and are all made good by it. Or they
could be read to mean that only those who are just by the grace
of Christ received:

> (the just man justices
> Keeps grace: that keeps all his goings graces;
> Acts in God's eye what in God's eye he is – Christ);

only in them is the Word incarnate in Christ really revealed.

And if all humans, and only humans are immortalised, this must be by a discrete act of divine resurrection of the dead, and it is so that all might be judged, and only good Christians, and those 'anonymous Christians' who can plead insurmountable ignorance of the source of their goodness in the grace of Christ, could be made eternally good.

That the second reading there is the correct one is suggested by what can only be called Hopkins' exclusivism. It is made clear *passim* in his diaries, letters and sermons that, more even than a 'one and only one true religion' man, he is a 'one and only one true church' man. His sermon for Nov 23, 1879, for example, has Jesus, who in actual historical fact never knew he was founding a new church at all, founding none else than 'the Roman Catholic Church'! But it is on his poetry once again that this reading is secured as the more likely. For brevity of argument sake, take two poems on Mary.

In *The Blessed Virgin Compared To The Air We Breathe* the air is so imagined as to make up the full connotation of the archaic term, spirit, as in the spirit brooding on the abyss in Genesis: 'Wild air, world mothering air.' Yet the comparison of Mary to this creator spirit which breathes life into all living things, thereby giving birth to all, is comprised in that attribute of Mary whereby, as Mother of Christ, she is mediatrix, dispenser of all the graces that create and sustain a spiritual life, a supernatural life. ('She holds high motherhood/Towards all our ghostly good'.) And so, forming such a life in us above and beyond a natural life and existence created as an integral part of all creation, she gives birth to other Christs in us. Now here is a life beyond the one we experience, it would appear – 'Though much the mystery how/not flesh but spirit now.'

Similarly in *The May Magnificat* where the question he sets himself is why May is Mary's month, and the answer again is delivered in the same by now familiar creation terminology, in a long and powerful evocation of what is then summed up as 'This ecstasy all through mothering earth' that characterises spring. The answer in the end turns out to be: May, the festival time of the giving birth to all things, is Mary's month merely because it reminds her that she gave birth to Christ, 'to God who was her salvation'. Once again the creative mothering that goes on in and as the natural earth is compared to – reminiscent of – another mothering that goes on in a supernatural realm.

Some later Christianised cultures, like early Irish culture, symbolised continuous divine creation through their ritual celebration of four seasonal feasts stretched over each and every year: *Samhain* on November 1, the beginning in darkness; *Imbolc* on February 1, the first burgeoning of life from the dead, dark earth; *Beltine* on May 1, the waxing of sun and growth; and *Lughnasa* on August I, fulfilment fruit, abundant life. With different gods for each festival posing no greater threat to monotheism, by the way, than the three persons in one Christian God with different jobs to do, or different parts of one great job. The newly Christianised Irish then moved the creator goddess of Imbolc, Brigid, into the divine hierarchy as a great saint, with all the attributes of the creator goddess intact, and called her 'the Mary of the Gael'. The indigenous peoples of Latin America when Christianised achieved a similar result by simply transferring to Mary herself the attributes of their creator goddess, Pachamama; and one of their recent theologians talks of the Holy Spirit becoming incarnate in Mary as the Word was becoming incarnate in her son.

The point here is that both the early Irish and later Latin American aborigines thought of earthly, and particularly human life in both its temporal and potentially eternal dimensions as a function of one and the same continuous divine creation in which creatures participate – forever forming, reforming, transforming according to the same natural process initiated and sustained by a non-natural (supernatural, preternatural) entity. And Mary and Brigid, when attached to particular ancient creation festivals, are thereby depicted as creators more like the ur-creator with whom they collaborate, than other creaturely co-creators are. So no dualism here, then, of a natural life and an eternal life, the latter the result of special, additional acts of resurrection or recreation 'after' death. Some ancient Irish Christian texts, like the *In Tenga Bithnua*, seem to refuse even the Platonic dualism entailed in the vision of God conferring eternal life on a thoroughly spiritualised humanity, suggesting instead that the transformation through death, impossible as it is to envisage during this life, would yet be continuous with all preceding continuous creation (evolution or transformation), and so result in a transformed natural world of which we could still be both integral part and immanent power, while more in God and God in us, as much as that is possible while the essential distinctions yet remain. (Perhaps Hopkins should have learned to read old and medieval Irish, or even read more Hiberno-Latin, instead of Welsh.)

Leaving aside issues of strict, not to say narrow orthodoxy, these ancient and continuing Irish ways of dealing with Mary require no dualist divine fashioning and furthering of life, either from the beginning or for the sake of immortality – the one natural, the other supernatural. And so neither was there required a dual act of perception and formulation, the one envisioning the whole natural creation of which we are so integral a part, its ultimate depths and furthest prospects; the other attempting to depict some life and existence of which we know only through the propagation of doctrine. For poetry can do nothing for the latter, the propagation of especially revealed doctrine, other than add to its linguistic expression what will now look rather like an artificial gilding of innovative rhythm and stress, sound and rhyme and pattern of word sequence. Whereas for the former, these very same innovations can enable the deepest and richest inscapes of reality to become manifest; can form a clearing in the forest, as Heidegger would say, in which Being can come to light. And since that is the proper business of poetry, the peerless inscapes of poetic vision and the corresponding instress of resulting revelations, a poetic corpus will be the poorer for the presence of a natural-supernatural dualism at its heart and core.

PART FOUR

The Creation as a Moral Project

CHAPTER SEVEN

Moral Gods and Tsunamis

Two statements, from God, through the lips of God's prophets.
The first through the lips of Isaiah (45:7, 18):
Forming light and creating darkness, making shalom (meaning not merely peace, but well-being, prosperity of life) and creating evil, I am the Lord who does all of these things' ... Thus says the Lord who created the heavens, who formed the earth and made it.

The second through the lips of Ezekiel (28:1-13)
Thus says the Lord God: '(you were the very model of perfection ... on the day you were created, but) because your heart is proud, and you have said, "I am a god" ... yet you are but a man, and no god, though you consider yourself as wise as a god ... (because) by your wisdom and understanding you have gotten wealth for yourself ... by your great wisdom in trade you have increased your wealth ... because in your widespread trading you were filled with violence against others ... because you consider yourself wise as a god, therefore, look you, I will bring strangers upon you, the most terrible of the nations; and they shall draw their swords (strap on their suicide bombs?) against the beauty of your wisdom and defile your splendour ... (and) you shall die the death of the uncircumcised by the hands of foreigners; for I have spoken, says the Lord God.'

Now we may possibly need no more than these two statements in order to realise that, according to the Christian Bible, God the creator, and as the creator, takes full and immediate responsibility for virtually everything in our world that we would consider an evil. Full and immediate responsibility, that is to say, for natural disasters as well as for all the evils that human beings do to each other, and to the lean earth that nurtures the lives of all, whether through military or economic strategies, both in despoiling each other of life and livelihood, and in coming to kill our despoilers in turn. Just as the same creator of the cosmos takes equally full and immediate responsibility for all that is good for us and for all creatures, in the pouring out of existence and life without stint and with no definable limit. 'Making shalom and creating evil, I am the Lord who does all of these things.' 'I will bring strangers upon you, and despoil your splendour, and you will die the death, for I have spoken, says the Lord God.'

Faced then as 'people of the Book,' as Muslims call us – a Book or Bible, incidentally, that they too regard as depository of divine revelation – faced with sombre reflection upon the death and devastation visited with the recent tsunami on hundreds of thousands of our fellow human beings, we are directed by these two quotations to look again at what we know of God the creator from creation, as we seek for any answers that may be available to that anguished question that the sight of death and devastation and suffering on such a vast scale forced so peremptorily to the surface of the mind that even the most superficial of the news media echoed it round the whole world. For, of course, a similar anguished question can arise equally out of the single death and suffering of any loved one, and that kind of question is then always virtually present. Yet it is usually kept dormant as much as possible beneath the noise of the busy daily traffic of our minds, until the passing roar of the tsunami forces it once more to the surface. And then as the anguished question is so suddenly forced to the surface, like the broken bodies spat out by the swollen sea, and religious folk are naturally the first to be asked urgently for an answer, it is little wonder that we come up with less than adequate answers; and in the case of Christians in particular, with less even than the best answer that the Bible makes available to us.

Take a brief look at some examples of the answers most commonly given. But first take a look at the way in which the anguished question is most frequently formulated, if only because one of the best ways of ensuring a false or inadequate answer to a puzzle about any feature of existence, is to pose the wrong question to it. The question made urgent and unavoidable by this appallingly destructive tsunami is most often formulated as follows: where was the loving and all-powerful God when the tsunami took its awesome toll of human life and livelihood; and how could God allow such a thing to happen in what so many suppose to be his creation? Now it should take but a little thought to realise that this is the wrong question, for it is based upon a very inadequate, if commonly assumed idea of divine creation, one that may be briefly outlined as follows.

We think – and too often teach – about the creation of the world as if it referred to a one-off act at the beginning of time, a one-off act that put the world there in the first place or, literally, as the first place. And after that first moment of the existence of space-time (as our scientists now call our universe), God did no more creating. Instead, God intervenes in the completed creation, now and then, to remonstrate, to punish, to repair, and eventually perhaps to put an end to it altogether, as a beginning had been made to it long, long ago. And on that flawed and un-Biblical assumption about creation, we come to think of God habitually 'squatting outside the world', as Marx once put it, keeping a very close eye on the world, of course, but now as one of our fancy theological formulae has it, using his 'permissive will' rather than his creative will. Just *permitting* everything that goes on in nature and history, except on those rare occasions, each once again a one-off in itself, on which God does intervene, either to forbid some creature to do some damage or to continue to do damage, or indeed to encourage some destructive natural process in nature and history to do some damage, as a punishment perhaps for rebellious creatures like ourselves.

Now that double-barrel question: where was God during the raging destruction of the tsunami; why did God allow it to occur? shares with the first answer to be given the assumption that God was not here in the creation at the time, or not in the Asian sector of the creation at any rate. For the first theological answer to the question is that God's creative will was not active in the tsunami. God did not cause it, but merely allowed it to happen. As if permitting something which one could prevent exonerated one, while one could not be exonerated in the case of an active, hands-on causing of that something. Clearly a piece of plain, see-through nonsense, and as un-Biblical (as the two opening quotations on their own should show), as is the picture of the habitually absent, if frequently interventionist God, which both question and answer assume.

Then a second and common kind of answer, trying to block the second barrel of the question: why did God allow it, wherever

God may have been detained at the time?, simply and humbly admits that it is a mystery. For the ways of the infinite and unfathomable Creator God are mysterious by nature, and far beyond the comprehension of the human mind. We do not therefore know, we simply do not know, why God did not intervene in order to prevent or at the very least curtail this particularly destructive tsunami. But we cling to our faith and trust that the Creator of this otherwise wonderful world and the giver of this otherwise wonderful life, will somehow, somewhere, sometime restore that life to all who were choked or suffocated or battered to death, or destitute and bereaved, in the course of that terrifying onslaught. A future divine intervention would counteract the lack of intervention on the occasion of the tsunami in the Indian Ocean. And in the meantime, we reciprocate the steadfast love of God that gave us this otherwise wonderful world and this otherwise wonderful life within it. We reciprocate by obeying God's command to us, by loving our neighbour as ourselves, and by doing everything in our power to soften the sorrows of the survivors with such spiritual and material support as may help them to rebuild their lives and their livelihoods. For by such steadfast, practical love we may enable them to feel again that limitless hope for life that the world, when it is good to us, holds out – and hope is perhaps the most precious commodity of all.

Let it be said straightaway that one should never attempt to do anything so crass as to cast this answer aside. Because the people to whom it is the only answer they can bring themselves to give, and who live by the faith it still leaves in their hearts, are surely amongst the noblest people we are ever likely to encounter in this motley crew that makes up the human race. Yet there are three things about it that should help rather than hinder those who give and live by this answer. First, God's ways with the world are not entirely mysterious. If we follow the direction of our opening biblical quotations, we may find that we know more about God's ways with the creation, continually creating both good and evil, and that may help us to reformulate the question away from the presumption of divine absence, and answer it more adequately, or at least to the point of not having to say: mystery, mystery, and little or nothing else. Second, some of our Christian brethren are so far from finding it all a mystery, that they know precisely why God allows such devastation of human lives: God is punishing us for our sins. And third, and consequently, our secular humanist critics can claim that a god who intervenes, or does not intervene as the case may be, in this manner does not deserve a single shred of allegiance, and is merely a figment of the human imagination gone monstrously wrong.

That last is the response of some prominent Irish journalists such as Patsy McGarry of *The Irish Times* and Vincent Brown of *The Village*. The God who allows the carnage of the tsunami is not the God of love that Christians claim to be the Creator, for such a God could not be exonerated from failing to intervene to prevent that pitiless crime against humanity; not by any more gracious intervention he might then make thereafter, nor by any shabby pretence that all that he was doing there was allowing the tsunami to punish the sins of humankind – of innocent children also? Brown concludes his piece decisively: 'Our rationality and our Bible tell us there is no loving, all powerful God who intervenes in our lives, who cares for us, on whom we can depend.' McGarry concludes his piece with a more positive secular humanist type of credo: 'All there is to this existence is the still, frequently sad music of humanity. Let's look after each other, as no one else will.'

John Waters, also in *The Irish Times*, did recommend to Brown and McGarry that they might consider an alternative position. A loving and all-powerful interventionist God might very well have its very existence contested as a result of the recent tsunami, and a multitude of other instances of such devastation wrought upon innocent people in this world. But what of a world in which God has to struggle, time and time again, against destructive, adversarial, that is to say, satanic forces within it? Why could we not accept the existence of a divinity on these terms, as some Jewish and Christian theologians have suggested? Why not place one's trust in a totally good and well-intentioned God but, as Woody Allen once put it, a God that looks at the moment very much like an under-achiever, and a very long way from omnipotence? To which the answer must be: what reason would we have to put all of our faith and trust in such a creator God and, joining his still uncertain struggle, spend our lives, at times quite literally, in helping others to life and life ever more abundant – instead of joining the old *carpe diem* cause and grabbing all we can for ourselves, however dire the implications for others and for the very planet we daily pillage? If God is still struggling with adversarial forces, then these are either in the world because he created them in the same way as he created all other parts and elements of the world, or they exist independently of his creative power; and in either case what reason is there to believe that he could succeed in the future, or in some other world?

Indeed, if there is some other world in which God does not have to struggle against some adversarial forces, why could he not have created the other world in the first place, and left it at that? And there is really no point in trying to mollify our mistrust by saying that God suffers with us in the lack of success that characterises these struggles. For that is to do nothing much more than to add one more sufferer to the exponentially increasing toll; and the fact that he is the biggest and best sufferer of all, increases the problem rather than make it a problem of more manageable proportions.

There is no denying, then, that secular humanists have much success in criticising these, the most common Christian solutions to what is more generally known as the problem of evil, a problem that the Asian tsunami posed once more in the most unavoidable of terms. There is point to the humanist claim that with these answers Christians are surreptitiously skirting the problem while gradually moving away from it, and covering their tracks with closing and copious applications of much blether about love. Yet before seeking a fuller solution, along the lines of our opening quotations from the Bible, it is only fair and proper to cast a brief critical eye on the secular humanist case itself in its fuller form. For that case might well escape unscathed if it was left where Vincent Brown leaves it, with the insistence on the non-existence of a loving, all-powerful Creator as the best excuse by a long shot for his apparent failure to show up on the occasion of the tsunami. But Patsy McGarry goes much further than that, and in two particular directions. First, with his suggestion that the problem of evil is a problem self-made by, and only for believers and their theologians. And second, with the concise little bitter-sweet credo with which he ends his piece: 'All there is to this existence is the still, frequently sad music of humanity', and all the help we may expect in our worst troubles will come only from our fellow humans.

McGarry certainly represents the stance of many secular

humanists in appearing to pretend that the evil in the world, an evil that comes in so many forms, that marches through all of history on its indomitable way, increasing in some forms as it begins to be contained in others, an evil that causes so much unnecessary death and devastation, disease and pain to the human species, that that evil is a problem only for believers in a loving God. It is not a problem for which some solution, or at the very least some cogent explanation, needs to be sought by non-believers and for non-believers. It is not a problem for secular humanists; it is simply in McGarry's immortal words, just 'life, as lived'. Even though much, if not most of the cruellest and most gratuitous death and devastation dealt out to humankind – not to mention to the rest of life on this earth – is daily dealt out by humanity itself, the very humanity to which McGarry directs his closing little petitionary prayer to help each other. At this time of the tsunami, the most powerful and best endowed nation, the USA with its little ally, Britain, is indeed engaged in saving us from what it has defined, if not invented, as international terrorism, and in the process killing as many innocent people and leaving as many suffering survivors, as the tsunami did and does. Just as in the last world war the very same allies deliberately massacred innocent civilians similarly numbered in their hundreds of thousands in Hiroshima, Nagasaki, Dresden ... Indeed a much more balanced piece on the tsunami, also in *The Irish Times*, by Fintan O'Toole, outlined the amount of responsibility for the human devastation it caused and that lay at the door of human impoverishment and neglect of the condition of their fellow humans in the devastated areas.

All of which makes one ask all the more insistently: by what fine tuning of his hearing apparatus does McGarry hear only the still, frequently sad music of humanity, and tune out the constant cacophony of whining shells and exploding bombs, the screams and cries of survivors, and the muted moans of all the hundreds of thousands of starving and stunted, for which we and our fellow humans are to blame? And by what scrambling of the neural connections of his brain that look after logic can he conclude that such awesome, persistent and apparently unending images of man's inhumanity to man causes no urgent problem for secular humanists, no peremptory demand for an explanation? The sheer shallowness, not to say evasiveness of the secular humanist position at this point is simply self-evident.

Back to the opening quotations then, and to the clues they offer when, in the relevant contexts, they present the God who claims to create evil and to make war, as well as to create that final peaceful prospering of life called shalom, when they present the God who does all of these things as God the Creator. These clues point directly to the perception that is then verified, and in a number of different ways, throughout the Bible. This is the perception that God as creator is the universal cause, and as such takes responsibility for everything and every process and every event that go to make up the nature and history of this cosmos, good, bad and indifferent. The perception that it is God precisely as creator who takes responsibility, though not sole responsibility, for everything that is or happens, can be verified by the Bible reader in a number of ways. For instance, by noticing that the Bible counts as creation features of divine activity that we are accustomed to file under other headings, such as salvation and consummation. It follows further from this perception of divine creation that, although from God's point of view creation takes place in some eternal presence, from our worm's-eye point of view within space-time it is perceived as an act of creation continuing through time and space; it is continuous creation. For God is creatively at work in all of the entities that make up the evolving universe at all times and places, forming, transforming and empowering their natures and behaviour.

It follows further from this perception of continuous divine creation, that the contrast between divine immanence and divine transcendence as normally understood and deployed, as the contrast of God within the universe and God squatting outside the universe, falls away and must be replaced, or re-interpreted by the preferred Biblical imagery of double immanence. God is within, working within all creatures, closer to us than the sandals on our feet. And at the same time, as Paul agreed with the Athenians when he tried to introduce Jesus to them, in God, as their poets put it, we live and move and have our being. There is therefore no question of God completing the creation in six days, or at Planck time 10^{-43} of a second into the Big Bang, and leaving it thereafter to work according to the natural laws inscribed in it – then going back to squat outside the world, and returning occasionally to tweak it in one way or another, for one reason or another. There is no question of an interventionist God in the sense commonly understood when the question is formulated: where was this Creator God when the tsunami devastated the coastal populations round the Indian Ocean? When the question is put in that form, it carries a presumption of absence, as in the question to a bereaved mother: where were you then when your little one was playing with the cooker and sustained fatal burns?

As even the *Penny Catechism* knew, God the Creator is present and active everywhere – which prompted the classroom joke about the little boy who was heard to utter the anxious but perfectly orthodox riposte, 'I hope he's not in my pocket eating my sweeties.' Therefore God's power and presence was in the tsunami as it raced across the ocean to kill before the coastal populations had a chance to escape. God was in the mud it churned up to choke the lives out of little children; in the crashing debris it swept along to crush their small bodies; in the broken bones and gaping wounds of those it left half-dead in its wake; in the bewildered anguish of those it left bereft of their loved ones; in the people who rushed to their aid, to enable them at least to bury their dead with some dignity, and to bring them clean water and rice, and tents to shelter them; in the monsoon rains that then came to make their plight worse again; in the donations of Irish school children to buy a boat for a fishing village that lost its whole livelihood; in the child-traffickers that came to prey on the children that had survived; in the sun that rose beautifully over a calm sea the following morning, to display the utter indifference of the cosmos, or to bring some rays of warmth and hope, depending on how this one or that could take it. In short, in the pithy words of the *Penny Catechism*: God is everywhere, everywhere. 'Forming the light and creating darkness; making shalom and creating evil... thus says the Lord who created the heavens, who formed the earth and made it ... I am the Lord and there is no other.'

Following further, then, the clues from our opening quotations, we are directed to the proper formulation of the question raised by the tsunami, and to the source from which an answer may be derived. The right question is not: where was God when the tsunami struck? But rather: what on earth is one to think of a divine creator whose presence and power is undeniably implicated – yes, in all good things of course – but now, most specifically, in the most destructive events and processes, natural or wilful, that cause us poor mortals such loss and suffering? And in our attempts to answer that question we are directed to what we can know of the ways of God the creator from the creation itself of which we are such an integral part. So that we cannot hoist the *'It is all a mystery'* placard from the very outset, for we are bound to know something of the ways of a creator from the ensuing creation. And to the anticipated objection of some Christians, that we should be directed by biblical quotations to the Bible rather than to the creation in search of answers, one can only reply that in the Bible itself, and with a frequency that perfunctory readers of the Bible might be surprised to discover, we are directed to that same creation when being asked to accept certain strong biblical views of the ways of God with humans, and vice versa.

For example, when Jesus wanted our ways with each other to be made to correspond to God's ways with the world, so that we should be perfect as our heavenly Father is perfect; when he wanted us never to return evil for evil, but rather to do good always to those who perpetrate evil against us, he pointed to God's ways with the world, ways that we can easily verify from the commonest of worldly experience. He pointed out that God makes the sun to shine on, to warm and to light the way for the good and the wicked alike, and God refreshes with his rain the just and the unjust. In short, he pointed to God's good gifts of water and fire, sources and conditions of all burgeoning life on earth, and asked us simply to note that God offers these and all of the life that they make possible, to all without distinction of persons, to the good and the bad, without condition, and without waiting for penance or even repentance. Of course, Jesus did not mention on this occasion that the sun's heat can increase and turn fertile land to arid desert, or that tsunamis are water also, but now in one of its most destructive modes. Nevertheless, he and the Bible point us to what we can know of the Creator from the creation as our means of formulating the right question and arriving at whatever answer that proves to be within our human reach. So we can go with confidence to the Bible and simultaneously to our experience and knowledge of the creation, including the increasing knowledge of the ways of the creation garnered by our scientists, in search of an answer to the question we are now forced to ask, properly formulated at last. What follows is a brief outline of the kind of answer we may find from a combination of Bible and science, and just plain human experience of this creation.

According to the opening of Genesis and the opening of the Gospel of John, God creates the world by forming a word. This word is not to be thought of as an expletive, a command – light, be! – but rather as an intelligible formulation, a formula, most likely a mathematical formula of the kind that occurs in the laws of physics, like E=MC2, by which we know something of how things are formed and transformed (in this case energy into mass, and vice versa). God creates therefore by means of such designs or plans. Furthermore, Genesis already has the notion that these forms or designs produced by God are not static forms resulting in a world made up of static natures of things, and so subject to immutable natural laws. On the contrary, the poets, the singers of tales who made and told these ancient myths, described these forms in their own inimitable way as dynamic forms, that is to say, forms that generate other forms. So God is pictured forming the land and the sea, and then commissioning these to form in turn: the earth must generate flora and fauna and creepy crawly things; and the sea the great sea monsters and all things that swim; and it is hard to tell whether the birds were brought forth by land or sea, or directly by God. And then these forms of living things are commissioned to propagate after their kinds. In this image of dynamic forms being created such as to bring forth and propagate other forms, the first intimation is given of the fact that what from God's side is creation is seen from the creature's side as evolution; and it is otherwise evidenced in that ancient world both in myth, and in the earliest philosophy – think only of Heracleitus, and his early followers, the Stoics.

Now, it hardly needs saying, such embryonic images of the evolution of life are light years away from the exponential advances in our knowledge of these cosmic matters that is still accelerating, as we speak, in the science of the modem era. But there is this in common between the primitive and the highly sophisticated here: modem scientific cosmologists now reckon that the primordial formulae, the ones that rule the whole creation-evolution of the cosmos through various emergent local versions, are of the following specification. They are formulae, designs, plans or structures that by sheer repetition and combinations produce new forms – in short, truly creative forms. At least one genius has even produced a computer model of such a formula, in an infinitely simplified form, of course, and you can download it from the internet and run it and watch it repeating until it gives rise to new forms or shapes. Now these formulae, designs or plans are clearly mind-like entities, or they represent a mind-like entity in action. And although people from both sides of the unfortunate religion and science divide want to challenge this conclusion, each driven by their own proprietorial interests – some religious folk thinking that they have a monopoly of knowledge of God; some scientists insisting that what they are coming to know is not a mind, and certainly not the mind of God – the conclusion is too obvious to be denied, namely, that in seeing something of this creative-evolutionary forming process, we are seeing something of the mind of God. This is accepted by as many scientists as reject it, as also by Christians from Jesus onward, who recognise the fact that we can know of the Creator's presence and power and ways with the world, from our experience and knowledge of the creation.

Needless to say, both the sensible scientist and the humble believer must admit that what is known of the creative power and the creation in which it is revealed, represents a tiny shaft of light in a still mysterious darkness. Our scientists, for instance, reckon that the matter they so far manage to study in detail is but a fraction of the stuff of the universe; behind it lies a scarcely penetrable array of dark matter, or energy, or something. Yet such essential humility should not prevent us from contemplating this mysterious world to which our destiny is so obviously bound, and seeking as good an answer as we can now give to the correct question posed so forcefully once more by the tsunami.

So, we know that the Creator formulates, gives origin to forms, perhaps to a field or fields of energy with which space-time comes into being, and from which emerge the particles that make up the hard matter in which our current form of bodily existence is formed. All of these forms of existence are then both dynamic and interactive, so that they inform each other, and simultaneously transform themselves and each other, as they adapt and evolve according to the universal formula for all interdependent and co-operative forms of reality that thereby make up one universe. As the universe progresses along these emergent, creative-evolutionary lines, life emerges, self-propagating systems, that is to say, some of which are creatively evolved to the point at which they can be conscious of the information and transformation to which they, like all that make up this cosmos, owe their nature and existence. And latterly, at the point now reached on earth, the universe evolves to the point at which a species emerges that is self-described as *homo sapiens* – that means, humanity-the-wise ... dear, O dear ... what a self-description for what is arguably the stupidest species in the entire universe – a species so creatively evolved as to be reflectively conscious of the transformative formulae that constitute creation-evolution, and to be able therefore to deliberately modify all of these forms. In this way, and at one and the same time, we reach the status of a quasi-independent, if always derivative co-creator, and become conscious of the Creator who is continuous source of all creation, including our own creativity. Already at this point of its development, then, the reflective or self-consciousness of the human being includes the consciousness of the Infinite Self.

Indeed, it may well be, as people who plumb the more profound depths of this common religious experience tell us, that it is in God's reciprocal consciousness of our conscious selves, that the Creator enjoys a form or kind of self-consciousness otherwise unavailable. Whether that be true or not, the double immanence is now in place, by which we consciously live and move and have our being in God, and God consciously lives and acts in and through us. And the purpose of creation is similarly realised: the purpose is what the act of creation itself achieves, namely, existence and advancing life in well-being. No additional, ulterior motive is either necessary or welcome. Notice how the great Eastern religious traditions concentrate much less than we do on laying out detailed moral paths to eternal life. They concentrate rather more on the kind of practices that allow us to pass through the self-centred, self-consciousness of the human self, to a close encounter with Brahman, the Absolute Self. To them suffering, death and tsunamis seem almost incidental to their journey and its goal; whereas we with our moral obsessions cannot be like that. So two further features of divine and human creative evolution become central to our Christian philosophy.

First, destruction and, for living systems, death are integral and inevitable parts of the evolutionary process by which the divine Creator continually creates this world, and so advances the world towards the evolution of ever higher and more deathless forms of life. In a created, creative, evolving universe more complex and advanced forms of existence and life are created and evolve through a transformation that inevitably involves a deformation, a death of existing forms. Stars are quite literally burnt to death in order to give birth to carbon, the building material for living things. This violent process in the macrocosm is mirrored in the microcosm of the violence to a woman's delicate body at the birth of every new addition to our race. The cooled and hardened plates of earth that move over the molten matter at the centre, the weak points in these and the points at which they catch and snap, produce earthquakes and volcanoes, and send tsunamis racing across the oceans to visit violence on distant shores – all part of the process of burning and burning out that produces carbon and life. And with living things it is even more true that, both as species and individuals, they literally live off each other, at each other's expense, and correspondingly they give of their lives and livelihoods that their offspring and others may live.

Until finally, at the highest form of life we earthlings so far know, reflective consciousness, mind, knowing and loving spirit, is driven by the mind-like Creator of all dynamic and intelligible forms of existence, to evolve from animal to superseding species, until it appears to have reached the point with *homo sapiens* at which it can survive the disintegration of the grosser material forms in which it currently develops and matures, to live on perhaps in the forms of energy that pre-exist the particle forms of hard matter emerging from them; and from these fields of energy, live on in the creative force that is present within and forming and empowering all; to live on, in other words, in the eternal Creator, who is forever operative within the universe, transforming it through consecutive forms to life and ever higher forms of life. However the process of creating-evolving higher forms of life, making current forms obsolescent; however the process of bringing life from death is illustrated then, God is the author equally of life and death. The same power, the same benevolent force that brings life into being, brings with it an always intermittent dying and death. As the poet Dylan Thomas saw it:

The force that through the green fuse drives the flower
Drives my green age; that blasts the roots of trees
Is my destroyer
The force that drives the water through the rocks

Drives my red blood; that dries the mountain streams
Turns mine to wax.

But the pattern that appears is this: the ultimate creative power in the universe, conventionally called God, brings death into life only to bring out of death life higher in quantity and quality, until their appears to us inevitably, if in hope, the prospect of life in its most god-like state, conventionally called life eternal. For all we can see of creation-evolution, the deaths of things inanimate and animate, species or individuals, are an integral part of the process of *élan vital*, the advance of life to the most god-like form that created life can attain. And that in turn means that death in itself is not an evil thing or, at the very least, it is a necessary evil on the way to final shalom, the life of peace, eternal rest and well-being that is threatened no longer with death and its harbingers, suffering and sorrow and the fear that defeats hope.

The second feature of this divine-human creative-evolutionary process that needs to be noted is this: the most advanced species we know, *homo sapiens*, humanity-the-wise (God help us all) has in fact advanced in wisdom at least in this respect: in our ability to understand the continuous creative processes employed by God in the world, we have reached the point at which we are increasingly effective co-creators of life and life more abundant, and that admittedly amounts to considerable wisdom. However, and here is the catch, from the very origin of our species, we seem to have been, and still to be, congenitally tempted by the very power and prospects of that wisdom, to use it, not as God does in favour of all existing and living things, but rather in order to advance our lives and livelihoods at whatever cost in ecological terms or to other human beings. Our very wisdom tempts us to such self-serving, self-aggrandising projects, and the whole of our history witnesses to the fact that we all of us fall for that temptation. This, incidentally, is the true, biblical notion of the fall and original sin. The serpent in the Genesis story is an ancient symbol of wisdom, by which we are tempted and fall. (So we can forget about that nonsense that Augustine inflicted upon theology, for all that Pelagius could say or do, about some peculiar sin transmitted automatically by the very process of human generation, and inevitably staining every human neonate from the instant of its conception.) As Ezekiel put it to the king of Tyre: 'by your wisdom ... you have gotten wealth for yourself ... by your great wisdom in trade you have increased your wealth.' That wisdom and the higher standards of living it brings, tempts us and we fall.

It tempts whole United States and European Unions of us alike, and not just the atheists in these united nations, but the Christians as well. And we then, god-like (Ezekiel: 'because you think you are wise as a god') secure for ourselves all the supports and enhancements of life, and deprive others of these in the process. Whether we go to war for oil, or land, or other spoils, or even 'to spread freedom'; or use strategies more effective than warfare, economic strategies like acquiring the labour and natural resources of other countries for as little as possible; or using subsidies at home as effective trade barriers; or giving development aid to the poor countries in the form of loans instead of gifts, thereby keeping them in crippling debt and subject to our directions and ends; or, as is happening presently in the occupation of Iraq, the channelling of so much of America's money given for the re-building of a country destroyed by its army (in the wake of Saddam Hussein and United Nations sanctions), and the channelling of Iraqi oil revenues also, back into the profits of American companies like Halliburton and Bechtel, while the illegal occupiers of that devastated land keep hundreds of thousands of Iraqis unemployed; and generally and for our own selfish ends raping and polluting land and sea and atmosphere, the whole of the good earth that gives and supports life for all. The list is endless of the devastations of humanity and the excess of death-dealing delivered by the wisest of humans themselves. And the list must also include the manner in which the predators incite the pillaged to draw the sword, strap on the suicide bomb, hi-jack the plane, raise an army, and get back what was taken from them, and more, much more, in the same destructive fashion.

So much so that natural disasters, such as droughts and tsunamis and the like, are not at any time responsible for a fraction of those whose lives are taken and destroyed by the actions and omissions of *homo sapiens*, humanity-the-wise. Nor, in so many cases, would the death and destruction wrought by such natural disasters amount to a fraction of what they do amount to, were it not for human self-centredness and its consequent evils of act or omission. Famine and flood kill far more people where trees and other wealth are drained from poor countries, kept poor. And the same with the tsunami. The Pacific has warning systems not shared with the Indian Ocean, and as usual it was the poor, kept poor, whose homes and livelihoods were most fragile, who died and suffered in by far the greatest numbers. The tsunami devastation too was not just the result of an act of God, but of evil human acts of selfishness, injustice and downright depredation.

Now none of that lengthy and detailed attribution to humanity itself of the blame for so much of the devastation and death that humanity suffers takes us any closer, in and of itself, to an answer to the question as to what we must think of the Creator of a world in which such things happen. It certainly does not exonerate God from responsibility, nor indeed was it designed to do so. If only because God is at least as responsible as is humanity for the original sin of human kind and for its devastating consequences for innocent and guilty alike – although few of us can be altogether innocent of either active or passive involvement in this original sin of the race, enjoying as we do the good life of the Green Tiger, the fastest growing economy of the European Union, without much evidence of troublesome conscience concerning the expense to others outside the union, and always with the possibility of soothing such conscience when and if it does trouble us, by generous donations to the victims of the most recent natural disaster, the tsunami. But to be fair to God, our opening quotations have some biblical prophets proclaim God's words to the effect that the Creator accepts responsibility for all of it. Indeed, God could scarcely deny responsibility, for it is the divine Creator that forms and empowers the natures and natural activities of all creatures, including human nature and all of its behaviour, both good and bad.

Nor, in view of such shared responsibility, is it possible to talk of death and devastation, whether caused naturally or by wilful human act, as God's punishment for sin. For one thing, given the shared responsibility, God would have to punish himself as much as us, if divine punishment for sin came into the reckoning. But once again, if the prophetic voices in the Bible are to be heard and heeded, God applies no divine punishment for sin, over and above the death and devastation we bring upon ourselves as a natural consequence of our evil-doing. The prophet, Jesus, when asked if some poor man's misfortune was God's punishment for his or someone else's sin said, no; it was just an opportunity for the reign of God, a reign or rule which consists essentially and always in creating life and life more abundant for all, to manifest itself. By his words and actions on this occasion Jesus sundered the link we too often make between evil that befalls and divine punishment. On other occasions, when people suffering life-diminishing illness or injury are brought to him, Jesus informs them that their sins are forgiven before he heals them; again so that they would know that they are always already forgiven, and that the restoration and enhancement of life is God's only and distinctive response to the threats to life that come either from nature or from human sinning against each other.

Or, it might be put in another way, that the healing and further enhancement of life is in and of itself God's forgiveness to humans for whatever evils they are responsible for bringing on each other. A point of view copper-fastened by the ruling of Jesus that if we are to be perfect as our heavenly Father is perfect, and if we are to extend God's utterly benevolent reign on earth, then we must never return evil to punish the one who does evil to us – for in any case, that would merely keep the vicious circle of evil-doing, death and devastation accelerating through history; but we are to do good, and only good to those who do evil, just as God does. For all that God, the eternal Creator, does is to carry on creating. God does not punish with any punishment that we do not bring freely upon ourselves, and the natural deaths that creatures must die on the way to fuller forms of life, is not punishment, but passage to eternal life.

And still the correctly formulated question returns in another form. Why, in the name of omnipotence, could God not have created a world in which people did not have to die, and in which a people intelligent and free would do only good and no evil? A number of answers have been proposed for that question, and one is as good, or as bad as the next. But what appears from all such answers is this: that the world created with such specifications would not include us, or any creature like us – persons embodied in corruptible flesh who have yet the dignity and stature of being freely in charge of our own destiny. So any conceivable answer to a question so formulated would assume our non-existence, and is correspondingly of no possible interest to us. But if we stick to the question, why did God create us and our world as it is, with death and human evil-doing?, we do have enough light from this creation to allow an answer. For it is just possible for us to see that God could not have created us, as we now are in our current hard, atomic material form, without simultaneously creating space-time, and that involves limits of time and space on such bodily forms; in short, it involves the mortality of such forms. But by the same token, by creating us, forming us by means of dynamic forms, God can bring us to successive and higher forms of life, ultimately not as limited to space-time as we now are. As Yeats envisages it in a poem that uses the ancient symbol of fire for continuous creation and prays:

O sages standing in God's holy fire.
Come from the holy fire, perne in a gyre, ...
Consume my heart away; sick with desire
And fashioned to a dying animal
It knows not what it is; and gather me
Into the artifice of eternity.

It is equally possible for us to understand that God could not create us through this dynamic, forming evolutionary process, could not have brought us to this level of our personal and truly creative powers, and of our freedom with regard to what we quite literally make of ourselves and our world, without exposing us to the temptation to think that we could take over the business of creating our world for our own selfish ends, thereby threatening and wreaking death and devastation on our fellow humans and on the earth from which we came and to which we shall return. And from all that we see from the Bible, and the natural world to which it so often points us, it is surely possible for us to know that God never punishes us with any punishment for sin other than what our own sins bring upon ourselves. And if God does not do this to us in our present bodily form, there is simply no reason to believe that God will so treat us to judgement and punishment in any future life we may hope to see, in that place or, better, that state we call heaven. Hell's punishment is here and like Sartre said, it is created by other people. There are no recriminations in heaven.

That last remark is actually a quotation, and it reminds me to end by saying – and you probably will not thank me for this – that you could have learned all I have tried to propose to you through the cumbersome logic of theology, from the far more delightful and accessible insights of our best poets. For the poet is essentially the seer, and the best poets can see better than any theologian into the depths of reality where the divine lives and works; and on that criterion, the best of our poets after Yeats is Patrick Kavanagh who penned the lines: 'Heaven/is the generous impulse, is contented/with feeding praise to the good.' In one short poem, that with his usual quirkiness he entitled, *Miss Universe*, he saw and said all I have tried to reach in this long chapter:

I learned, I learned – when one might be inclined
To think, too late, too late, you cannot recover your losses –
I learned something of the nature of God's mind,
Not the abstract Creator but he who caresses
The daily and nightly earth; He who refuses
To take failure for an answer till again and again is worn.
Love is waiting for you, waiting for the violence that she chooses
From the tepidity of the common round beyond exhaustion or scorn.
What was once is still and there is no room for remorse,
There are no recriminations in Heaven. O the sensual throb
Of the explosive body, the tumultuous thighs!
Adown a summer lane comes Miss Universe
She whom no lecher's art can rob
Though she is not the virgin who was wise.

CHAPTER EIGHT

The People of God Working the Natural World
The Primary Authors of Christian Morality*

In the last years of the old millennium, the Roman Catholic Church's Synod of Bishops gathered in a special assembly from which there emerged a working paper entitled: *Jesus Christ Alive in His Church: Source of Hope for Europe*. The section of that paper entitled, 'Serving the Gospel of Hope', waxes most eloquent on the precise manner in which the church can hold out hope to the new and growing European community, to make that community of nations an ever better place to be. The authors of that paper felt no need to qualify that phrase, the church, whenever and wherever they used it. They did not feel it necessary to say which Christian church they had in mind, or if the phrase denoted the whole Christian community in the world, scattered across the various Christian denominations. This could simply be, of course, because emanating as it does from a synod of Roman Catholic bishops, it would be obvious to everyone that the church intended was the Roman Catholic Church. But it could also be because, as in the document on Ecumenism from the Council called Vatican II, the authors simply assume that the one true church of Christians in the world that could claim Jesus as its founder subsists in the Roman Catholic Church. So that this synod could speak for all Christians. Just as the Vatican II document on Ecumenism assumes from beginning to end that the unity with which Jesus himself endowed his church also already subsists in the Roman Catholic Church, and will therefore be fully realised once more as the other churches more or less rejoin it.

With such impressions of the church in question lodged firmly, if perhaps somewhat insidiously, in our minds, the bishops' paper then goes on to outline quite succinctly the source and nature of the love which, as any Christian could see, holds out

* A paper first delivered to the Irish Lay Catholic organisation, Pobal, in Dublin, 19 February 2000.

the best hope for the European community, and indeed for the whole world. This is the love of God made real in act from beginning to end of creation; a love consummated in Jesus the Christ, and poured out unstintingly through the Holy Spirit. *Bonum diffusivum sui:* goodness of its nature overflows its container and source. And this goodness of God's love-in-action, experienced by all in the creative gift of life itself, and in the gifts of all the supports and enhancements of life and life ever more abundant to all – this, if anything ever can, must enable and inspire its recipients to be good to all others, thus to form communities of love in action, a natural and dynamic extension of God's primordial love, and the surest source of hope of limitless life for all. (Working Paper, n 72) A succinct but adequate and inspiring summary, we might well agree, of a Christian claim to serve the developing European Community with its gospel of hope. And yet, no sooner do the bishops develop this theme than misgivings begin to arise in the mind of the reader. Misgivings about a certain narrowness, a certain and almost certainly crippling restrictiveness of the vision invoked. Misgivings not altogether unconnected with the impressions of the understanding of 'church' just now noted.

Consider the following sequence of argument as the bishops' working paper develops its central theme. 'Undoubtedly the *first way* to live the witness of charity is to be builders of communion *in the Christian community.*' The paper does immediately add that, 'the witness of love *also* extends *beyond* the confines of the ecclesial community'. (n 73; emphases mine) But the impression is already indelibly lodged in the reader's mind that the first beneficiaries of Christians' embodying and bodying forth the originating love of God, are other Christians – if not indeed other Roman Catholics. This is reminiscent of the dubious compliment that Christians record being paid to their early antecedents, 'See how these Christians love one another.' Not, as one might expect from even a peremptory perusal of Jesus's Sermon on the Mount, 'See how these Christians love all of us, even though we are so often their sworn enemies to the very death.' So that the community which the Christians' embodiment of the love of God is to build into a community of solidarity, of mutual love and reconciliation, is first and foremost the Christian community itself, if not just the Roman Catholic Church. Not first and foremost the European Community, or any of the national communities that make up that growing family, or indeed the whole community of nations that covers the whole world. But then of course, having built up the church into such an exemplary community of mutual love and harmony, that church, as the paper puts it, will become 'the primary element of stability and communion even in society', and destined thereby, the paper adds, to 'promote the construction of a unified society'. (n 73)

This impression of a prior ecclesial self-centredness inevitably entails a correspondingly restricted image of God's own interests and activities. And it leaves everyone with an impression of a lesser God, an impression then fully confirmed by the bishops' immediate appeal to the institution of the Eucharist, the central Christian ritual. 'The Church's communion has its centre in the Eucharist, the primary place of encounter with Christ and his people. This encounter round the table of the Lord gives rise to fellowship, the characteristic feature of the Christian community'. 'Which,' they add, almost as an afterthought, 'extends its beneficial influence to civil society.' So now the impression is fully formed and secure; the impression, namely, that the first task of Christians is to build up 'the church' as a community of mutual love, reconciliation and service, but with the expectation and indeed the intention that this should be the priming element for communion 'even in society', promoting in the European Union, as it ought, 'the construction of a unified society'.

I must admit, it was when I suddenly thought, not about what was being said here, but about the bishops whose Working Paper was saying it, that I realised how inadequate, and in fact how downright wrong-headed this development of the central theme of God's love forming community through Christians in the world had by this point become. The body of men now telling me that the encounter of Christ with his people, the encounter that gives rise to a community of mutual, practical, life-enhancing love and fellowship for all, was an encounter that took place primarily in the Eucharist – this was precisely the same body of men under their leader, their Pope, that would not allow the members of their church to share the Lord's table even with members of other Christian churches, not to mention all of those people in the wider human community who are as yet somehow to be influenced by the Eucharistic community towards a reconciled human fellowship that alone holds out hope for life more abundant for all. I hope I can be forgiven for wondering whether I was being treated here to a piece of perhaps pardonable ignorance, or to a quite gormless though no less haughty form of hypocrisy.

For the Irish bishops, at least, could scarcely be entirely ignorant of the contribution that divided Christian communities make to the mutual ignorance and suspicion of, and the judgemental attitudes to the other, that are themselves contributory factors to the hostility that still obtains between divided communities in Northern Ireland, and that then not surprisingly erupts into overt violence when triggered by many of the assorted factors that trigger such violence on regular occasions. No amount of declaration to the effect that this fighting and hostility in Northern Ireland is not a religious war, nor is it even really about religion as such – even if such declamation does always have a substantial element of truth about it – can hide from view the serious damage done to the community fellowship in civil society in Northern Ireland and places like it, by the sheer maintenance of such hard and exclusivist denominational divisions, particularly perhaps during the most formative years of the Northerners' formal education.

And if the Synod of Bishops is right – as it certainly is right – to insist that the creative and holding centre of the communion of Christians is in the Eucharist, and that it is from there that it extends its beneficiary influence to the whole of society, then the persistence of that same hierarchy in fencing off the Lord's table even from fellow Christians does direct and increasing damage in depriving the whole society in question of this supreme source of unifying grace and love. Perpetuating instead such social and political divisiveness as may coincide with denominational divisions. Interestingly enough, the most persistent offenders in this matter are the Roman Catholic hierarchy, on the one hand, and the most fundamentalist of evangelical Protestant churches, on the other.

The Roman Catholic hierarchy gives reasons for the ring-fencing of the Lord's table even from fellow Christians. But these, on careful analysis, turn out to range from the merely unconvincing to the entirely inadmissible. The commonest reason given from Roman Catholic refusal to break bread at the Lord's table with other Christians, is that we must not celebrate a unity that has not yet been achieved. But that flies directly in the face of the Synod's own acknowledgement of the fact that Eucharist is the divinely resourced source for growing and widening unity of agapeic fellowship amongst mutually hostile human beings, before it can become the celebration of such unity. And so the reason for not engaging immediately in what they disingenuously call inter-communion, when in fact it is plain and simple Eucharistic communion, is no better than a pathetic piece of evasion of our responsibilities as Roman Catholic Christians, and an especial example of our accumulating guilt in not facing up to these.

Things are worse still, at a time when modern conversations between Christian churches have revealed so much unity of understanding amongst them, and in particular so much agreement as to what Eucharist is and does. Furthermore, at a time when Roman Catholic objections to the validity of the 'orders' of other Eucharistic presidents can be shown to be themselves invalid, one has good reasons to suspect that it is the resistance of other churches to an acceptance of authority over them by the Roman Catholic hierarchy as presently constructed and defined, that is Rome's only remaining obstacle to full Eucharistic communion with other Christian churches. And if that suspicion is confirmed, as it probably can be both from experience of some prominent inter-church conversations and from Vatican II's document on Ecumenism, then this hierarchy is placing at the very centre of Christian faith and life a set of ecclesiastical governmental structures that have been developed by very human beings indeed over the centuries. For in making the acceptance and maintenance of these, and of a church characterised by these, the condition of breaking bread with other Christians, it is in effect placing these higher in the hierarchy of Christian belief and practice than Eucharist itself, the divinely resourced source of successful social unity and increasing agapeic fellowship, in the Christian community as a whole and in the broader society.

As a further consequence, the Roman Catholic Church then presents itself as an inward-looking institution in human society. Quite in line with the impression conveyed in the Synod's working paper that the church is there to build up its own hierarchically structured fellowship first, while then hoping to see some agapeic fellowship in the wider society as a spin-off in some largely unspecified manner. No amount of dressing up in the finery of the ancient Roman Empire, no use of titles of lord-ship from feudal times, no list of invitations to leaders of other churches to join in prayer or in opening the odd holy door, can hide the fact that an inexcusable obstacle is being placed by Rome in the way of the unifying of the whole Christian fellow-ship of the world, and of what should be the concomitant, not subsequent, uniting in fellowship of all peoples and all national families in the world.

What to do about this, here and now? What to do in order to bring closer to realisation the vision of the Christian community as a true Eucharistic community first and foremost, and only very, very secondarily and non-essentially a community that exhibits a variety of organisational structures? Well, to speak for Roman Catholics, if our hierarchical leaders will not push or pull the ecumenical movement out of the doldrums in which it has now for some time languished; if they still, to change metaphors, insist on constituting, as they have from the beginning, the cow on the line, then the plain people of God in that church must take up their responsibilities for being the church. Go, break bread with your fellow Christians in their various churches. Follow the fine recent example of our Irish President. Invite the other Christians to your Eucharistic tables, for these tables are yours just as much as they are the priests' who presently preside over them. You will find yourselves mostly welcome at other Eucharistic tables, and your return invitations welcome. And where entry is refused, and invitations rebuffed, you will know that those who rebuff and refuse at the Lord's table and in the Lord's name will inevitably answer for that to the same Lord who opened his table fellowship during his exemplary lifetime even to the outcasts and sinners of his age, without any prior conditions attached to their coming to sit at table with him.

Yet even those who are most convinced that the *laos*, the laity, the people of God are the church, might not think that they should, or indeed effectively could follow President Mary McAleese and this piece of advice on 'inter-communion'. And they might point out with good reason that, even if they did, and even if more and more lay people were by this means successful in moving forward once more an ecumenical movement at present stalled in a surfeit of largely symbolic gestures, even then they would succeed in spreading only to their fellow Christians the source-resourcing love of God the creator, a love that is at work in the whole of creation. Leaving the rest of society to hope that by some unexplained osmosis it might spread to them also. It would be possible to make a response to this position at this point by appealing in the second part of this paper to a much more radical and more biblically based portrait of the Eucharist as Christ instituted it, before a clerical caste monopolised the concessions connected with it. This would involve a theology of Eucharist centred, not on its sacrificial aspect as retribution for sins committed against a God of justice – for that, however valid it may well be, is but a part, and the lesser part of the life-giving relationship between God and humanity that Eucharist both symbolises and effects – but centred more on the meal aspect of the Eucharist: taking with thanks, blessing God for the bread and wine, symbols of life and all the supports and enhance-ments of life, which earth has so unstintingly given and human hands have so painstakingly made, and in an overflow of grati-tude breaking and pouring out and giving again to all who would accept the unconditioned invitation to receive. Eucharist as a sacrament of creation then, a ritual enactment of the love of God always shown to all in all of creation. And therefore open to all, and not just to the baptised, and certainly not just to the re-cently shriven. But that would require a detour on the way to the conclusion concerning the church and its laity for which there is simply not enough time on this occasion.[1]

So may we continue instead with some further consider-ations concerning the role of the laity, the people of God as such, as the authors of moral value in human society in the world as a whole; the moral values of love and the rest, which they are in-spired in the Eucharist to serve to each other and to all their fel-low humans, by accepting the very symbols of life itself from the hands of the God who gives it in unlimited generosity to be shared by all.

---

1. A more extensive coverage of the multiple symbolism of creation, of church, and of Eucharist as a natural, sacrificial life-sharing meal, can be found in my *Christianity and Creation*, New York and London: Continuum 2006, particularly in the chapter on 'Cult,' section on 'Eucharist.'

II

Begin this second part then with a quotation from another document that comes also from the Roman Catholic hierarchy, the Decree of Vatican II on The Apostolate of the Laity:

It is the task of the whole church to labour vigorously so that men may become capable of constructing the temporal order rightly and directing it to God through Christ. Her pastors must clearly state the principles concerning the purpose of creation and the use of temporal things, and must make available the moral and spiritual aids by which the temporal order can be restored in Christ.

The laity must take on the renewal of the temporal order as their own special obligation. Led by the light of the gospel and the mind of the church, and motivated by Christian love, let them act directly and definitively in the temporal sphere. As citizens they must co-operate with other citizens, using their own particular skills and acting on their own responsibility. Everywhere and in all things they must seek the justice characteristic of God's rule. The temporal order must be renewed in such a way that, without the slightest detriment to its own proper laws, it can be brought into conformity with the highest principles of the Christian life and adapted to the shifting circumstances of time, place and person. (n 7)

Now this text, and much more scattered references in the same vein, if it means anything at all, means to say that the envisaging of what is good, better and best for all in the whole of secular life, that is to say, the whole of life lived in this temporal world (space-time) – in marriage and family, the trades and professions, avocations of all kinds, the whole of industrial, economic and political life, in fact the whole of what we call living in this world, the whole of what passes for morality in this world – is the special prerogative of the people of God as a whole, who occupy all of these areas of the common endeavour to secure life and life more abundant for all.

'Love must be made real in act,' as the poet, T. S. Eliot put it, 'as desire unites with desired' (although he added rather sadly these lines on the plight of the unemployed: 'they have only their labour to give/and their labour is not required.') To channel the love of God, made real in the act of creating boundless life eternally for all, into society at large and into the whole world, this must be done, can only be done, by the whole people of God, rather than just the hierarchy, in all the areas of life and means of livelihood that constitute the secular world, as they work in all these areas for what will be good, better and best for all. Notice that in this context where the whole of human morality is at issue as the channel of God's creative love, there is no mention of building first the Christian community, with some later addition of something about the wider human society in the world. Nor should there be. For morality, the envisioning and promotion of moral values, is the common duty and right of all. It is the inalienable birthright of every human being. It is theirs from the very fact of their being human, and it is not conferrable therefore by any authority, civil or ecclesiastical. The kind of texts just quoted, although their contexts might sometimes suggest that 'the laity' bear this moral responsibility for themselves in their societies as a result of their baptism, must always be read in the light of the fact that humanity owns this inalienable responsibility from the very fact of being human.

So then what we are being told in the texts currently under review, is that the laity are the authors of morality in this world, as they channel to the world the love of God expressed in all creation and fully embodied, as Christians believe, in the life, death and destiny of Jesus of Nazareth. To quote again from the decree on the laity at Vatican II: 'The effort to infuse a Christian spirit into the mentality, customs, laws and structures of community, is so much the duty and responsibility of the laity that it can never be properly performed by others.' (n 7)

Fine words indeed; but fine words that continue to be belied by some practices of the government of the Roman Catholic Church. For the hierarchy continues to behave as if the envisaging and promotion of moral values was primarily a matter of framing and imposing laws. Then it took advantage of the fact that Vatican II was essentially a compromise council, in this respect in particular, that it set beside the model of the church as the people of God a totally authoritarian hierarchical model of the church, which God had not given but human hands had made. So that the hierarchy continues to monopolise the right and duty to decide upon the morality that should build up the human community in the world.

Worse still, in pursuit of that same monopoly, the hierarchy has seriously misled the faithful in matters of morals, particularly in the areas of marriage and family life. The principal examples of this are only too well known, and need only the briefest of mentions here. There is the almost forty-year old papal persistence in the false moral teaching that all use of artificial contraception is intrinsically sinful, and the increasing cowardice of those priests and even bishops who know this to be false, but fail to speak out against it – despite the fact that the vast majority of the good Catholic laity all over the world has long since exercised its own good practical moral judgement in this matter – has exercised, in short, the very prerogative which the decree on the laity from Vatican II so fulsomely acknowledged – and come to a quite different conclusion on the use of contraception of all kinds.

There is furthermore the oddity, to say the least, of the hierarchical teaching to the effect that any introduction of legislation enabling divorce on any grounds whatever would alter 'God's plan for marriage', and should therefore be anathema to all good Christians. Despite the fact that the particular understanding and definition of what constitutes valid marriage, which they call 'God's plan for marriage' cannot be traced back much further than the Middle Ages in Europe; that the Bible is at best ambiguous on the subject of divorce; and that provision for divorce for Christians has always been made, from the Bible down to the most recent recension of the Roman Catholic Church's own Code of Canon Law.

And finally, in the course of recent bitter and divisive debates on the issue of Irish law and abortion, the hierarchy once again sought to appear to be in possession of simple moral absolutes, as it sought to ensure a simple and permanent legal ban on abortion in this country. Despite the fact that traditional Catholic moral teaching allows medical procedures that would kill the foetus, if these were essential to save the life of the mother. Now these procedures are obviously abortifacient, and the whole complex of these procedures together with their natural outcomes, can then be refered to as abortion in normal medical and popular diction. But then the hierarchy, and some at least of the lawyers concerned, seem to want to deny that abortion occurs in such cases, and for this purpose they use what can only in this case be called the casuistry of 'the act of two effects'.

The act of two effects means that if you perform an act, say, a medical procedure by which you intend to save a mother's life, and that procedure results also in the death of the foetus – the second of the two effects – then, as long as you did not positively intend to kill the foetus, you cannot be accused of procuring an abortion. Now only so fine a piece of casuistry could blind you to the fact that, if you perform and intend to perform a procedure that you have good reason to know will have both effects, you certainly must be said to have intended both effects. 'I certainly and intentionally drove my car over the plaintiff's foot, m'lud, in seeking to squeeze my car into the one remaining parking place. But I intended only to park my car; I did not by any manner or means intend to break even a small bone in the plaintiff's foot.'

The result of using this kind of casuistry in order to be able to say that the church never can condone abortion in any set of circumstances is, as with all uses of casuistry, to twist language from its common usage, and to do a grave disservice to a serious public debate that continues to exercise many countries. Thus the hierarchy, together usually with a vociferous section of over-loyal laity, is again and again engaged in a desperate attempt to possess and promote simplistic moral absolutes. Instead of openly admitting that what Catholic moral teaching permits under the so-called 'act of two effects' – anybody who truly understands the complexity of moral reasoning will realise that all acts have numerous effects, seldom if ever just two – is what most medical and ordinary people would call abortion of the foetus in order to save the life of the mother. And then, instead of taking a positive and helpful part in the difficult matter of defining terms such as the nature or degree of the threat to the life of the mother, and of the risk to the foetus, and so on, that would make an abortifacient procedure morally acceptable, churchmen seem to prefer to engage in simple denunciation of all abortion, thereby running the risk of misleading the faithful in matters of morals once more, and of misleading them not least on the true content of the best of Catholic morality.

It may be objected that there is too much concentration here on the church's moralising on sex and related matters. And whereas there may be a case for complaining that the Roman Catholic Church in the recent past in particular devoted too much of its moralising to these, there is also the fact of its centuries' old tradition of broader social teaching, and more recent evidence still of a change of emphasis from sex to social justice. There is much truth in that assertion of change, and it deserves an ungrudging welcome. And yet, especially in the context of the argument of this piece, one dare not forget the criticism of that long and recently revived tradition of Catholic social moral teaching made recently by the Professor of Sociology at Maynooth. This criticism was to the effect that even here the moralising took the form of handing down from church government principles and precepts which the hierarchs seemed to assume had come to them, more or less ready-made, from a higher authority still. Suspicions, then, of a further usurpation of the rights and duties of the laity in these matters, which our texts from Vatican II so robustly acknowledged. A usurpation the dire consequences of which Ireland witnessed in the now infamous case of the mother-and-child welfare scheme, which a minister of health in the government sought to make a matter of state policy, only to have this worthy effort brought to nought by, amongst other obstacles, an exercise of high-handed hierarchical argument from principles of a rather abstract nature and uncertain applicability in the concreteness of social-historical circumstance.

It might also be objected that all or most of the above is but a thinly disguised exercise in the verbal bashing of the bishops. To which one can only respond that it is simply true to say that the hierarchy, especially in this Ireland of ours, have seriously misled and do continue to mislead the faithful in matters of morals, even though the same laity have risen to their own obligations and nullified thereby some of the harm done to the church, by the contraception issue at least. For recent hierarchical moral teaching has caused a very great deal of unnecessary and unjustified hurt and suffering to the most faithful Catholics in the church. It has also, in the course of this process, lost to that same church more of the whole church's moral authority, that could otherwise be so invaluable to society at large; and lost more of the allegiance to the church of more Irish Catholics than would ever have been lost by paedophile priests or sadistic nuns and brothers. The hierarchy will no doubt answer for all of this to the same Lord, whose table they ring-fence not only from others of his faithful followers, but from those lost ones that the Lord insisted enjoyed the first and unconditional invitation to break bread with him.

Yet none of this, and no amount of unfortunately necessary repetition of it, detracts from the office and duty of bishops, as officers of good order, to critically and judiciously supervise the Christian life of their local communities, in all aspects of faith and morals. But it is necessary also to recognise that such supervision is an entirely different matter from continuing to act as if the laity did not have its own inalienable rights and duties in the business of translating the Spirit of its Founder into those moral values and prescriptions that can transform, as the Vatican II text put it, 'the mentality, customs, laws and structures of the (wider) community'. For one thing, the legitimate supervisory role of bishops in matters of faith and morals is best conceived in terms of service rather than law. Jesus told his chosen leaders to serve rather than lord it over the rest of his followers. And the service the bishops can offer in this respect is best conceived in terms of the discernment of spirits: whether, that is to say, it is the Creator Spirit revealed to all in its continuous creation and incarnate, as Christians believe, especially in Jesus of Nazareth, that truly inspires and directs the practical moralising of the laity – or some other demonic spirit of greed or lust for power.

And for another thing, there must be acknowledgment of the possibility that at times the supervisors may fail to recognise the Creator Spirit at work, as they presently continue to fail in the matter of the laity's development of a morality of contraception. And as they will no doubt fail again and again, for as long as they do not recognise and respect in practice what Vatican II acknowledges in theory, namely, that the Spirit of Jesus moves the whole people of God, directly and in its own right, to join the whole human community in God's good creation in the task of making life in this world reflect as fully and as justly as possible all the love and the unstinting grace that God pours out daily in and as that same continuous creation of existence and life.

So then, at times when such recognition of and respect of the role of the laity in transforming human society is either ignored or indeed denied, and sometimes not so much by the hierarchy as by vociferous sections of the laity itself, it is time for the *maior et sanior pars* of the laity, not to go cap in hand to anyone for permission to play this role, but simply to get on with playing the role. Trembling with fear as they do so that they may make an even bigger mess of it than their clerical leaders have recently managed to do, but encouraged by their faith in the Spirit of Jesus still living and working through them, as through others also, to fashion the world into the image that Jesus, especially in his Sermon on the Mount, painted of the reign of the creative love of God in God's good world – the kingdom of God. (The kingdom of God is not to be confused, of course, with any of the Christian churches still operative in the world, nor indeed with all of them put together, even if they could all be put together, which looks increasingly unlikely.)

The laity and supervisory episcopals then play their complementary roles: the former as the primary authors of Christian morality, and the latter as discerning supervisors of that quintessentially lay activity and of its end products. Both must recognise that the Creator Spirit, through the continuous act of creation itself, both inspires by this primordial act of love all creatures who can respond with love, and directs that love towards its essential moral outcome in such co-operation with the divine act of creation as will honour the divine will to make the benefits of creation, namely, existence and life in increasing and eternal abundance, available to all.

All must recognise also that the divine creator that emerges in the *persona* of Creator Spirit, as also in the *persona* of the Creator Word, endows with this same inspiration to love and guides with this same light to the exercise of true moral living, every one, quite literally, who comes into this world, or has or will come into this world. As John in his gospel states the matter explicitly in the so-called Prologue to that gospel, when he says that the Word without whom is made nothing that is made, enlightens all who come into the world. So that when the followers of Jesus's claim that this Creator Spirit or Word so indwelt the man Jesus that that man's life, death and destiny was the most perfect expression possible in human form of the continued presence and action of the Creator God in this world, a claim that is made under the image of incarnation, and when these Christians, as they came to be called, then claim further still that this Creator Spirit that was so incarnate in Jesus is then breathed into them by the very process of their becoming followers of his, so that that same Creator Spirit breathes through them into the world that same divine inspiration to love and the same light that shows the direction to the realisation of what is best for all of this world, there are a number of things that they must immediately realise.

First, they must realise that the same Creator who is personified as Word and Spirit does not morally inspire and enlighten

the followers of Jesus alone. Far from it, since the divine inspiration and enlightenment takes the fundamental form of the continuous and gracious act of giving existence and life unstintingly
to all in eternal abundance. Second, and consequently, that their
claim to be sources of inspiring love and enlightened moral
direction precisely as followers of Jesus baptised into his Spirit,
will depend just as precisely upon the degree to which they actually embody and body forth that Spirit – rather than do, what
to one degree or another they have always done in the course of
the history of Christianity, that is, to modify or even to betray
the Spirit of Jesus for the sake of their own weakness, or greed,
or lust for power. For example, as Christians of the first couple
of centuries seemed to know, but as quickly forgot as soon as
church and state cosied up together, any Christian who engages
in warfare betrays the Spirit of Jesus in the most complete manner possible. And, finally, we must realise, that the Roman
Catholic hierarchy is not the only, or even the main judicial arbiter of what is true love or true morals in this world; nor is the
leadership of any other Christian church; nor the leaderships of
all of them put together. All members of our race everywhere, as
recipients of the Creator's inspiration and enlightenment through
the creation itself, are by that very fact arbiters of good and bad
moral behaviour, and this whether they be members of other
religions, or of none.

What makes one group a better arbiter than others in such
matters is precisely the degree to which they live up to the
inspiration and enlightenment that comes to them in and
through the creation itself, or through whatever person or event
in the history of creation they say it has come to them especially,
and with what they feel is some consummate power and clarity.

Such considerations give its proper depth and its proper
extent to the claim that it is the laity, that is to say, the whole
community of people working in the world, that constitutes the
primary authorship of morality. So that there is no need to call
upon the laity to exercise their authorship, except perhaps when
their leadership in church or state are allowed to usurp that authorship and, worse still, then make a bad job of it. And even
then there is seldom if ever a need for such calls to be made. For
it emerges that the people are always in fact doing the job, however occasionally they fail to do it or do it right. The instance of
the papal ruling on contraception provides an example of the

laity going about their business in this matter, quietly and without engaging in any controversy with the hierarchy. For the primary means of creating and re-creating moral values and infusing them into the world at large is simply the living of our lives, the making of a living or, as some of our younger members would undoubtedly phrase it, the getting of a life, creatively, together in this evolving universe, in which animals and indeed all other entities have a right to well-being as well as we humans have. To be more specific, the primary authors of morality, and hence the primary authorities on morality, are the practitioners in every walk of life, in family and marketplace, in industry and commerce, in the professions and public life. Each knowing best that way of life, will also know best what is also the most just and moral way of conducting themselves in the course of that livelihood.

Yet a special and important part of the authorship of morality will always consist in regular reflective clarification, that is to say in this context, in the codification of the values which we, all of us, constantly and creatively evolve and infuse into this ever evolving universe, in the light of our birthright and our inalienable responsibility for each other and for morality itself. This codification takes many forms in those organised societies we call nations or nation states, and in the higher unions of these to which increasingly these now belong. It takes the form, for instance, of the policies which we elect and authorise our governments to enact. Most obviously, it takes the form of legislation by state governments, and this codification of moral value in the form of laws must also be subject to constant revision.

The implication of this point about constant critical clarification and codification of moral values in the case of the Christian laity in any church must be obvious. But their primary and essential part in this moralising process, so robustly defended in Vatican II, cannot be exercised effectively by waiting for such action from the laity of any church as a whole. Only a mistake as huge and as vital to their lives as the hierarchical decree on contraception is likely to galvanise the whole laity of a church into action; and moral mistakes of that magnitude by the hierarchy are, happily, not frequent. So representative bodies of the laity are necessary in order that this primary and essential responsibility of theirs be exercised regularly and in a wieldy and effective fashion. And Pobal would seem to be an obvious organisation to engage in this necessary task.

One last point: just to indicate the fuller context that talks like this one must seek in the theology of creation, of Eucharist and of church, the multi-layered context in which most of what has just been said would find its fuller explanation and more persuasive rationale.

Consider the Eucharist as primarily the sacrament of creation. Then in the breaking to each other of bread and the pouring out to each other of the wine, we celebrate our co-creative responsibility-in-action in God's good creation. As, true now to what we just symbolised, we break open our lives to enhance the lives of our fellow humans and of all creatures great and small. Sharing the staff of life, food and drink, and empowered to do so as the continuously creative God pours out life liberally to all, and all the supports and enhancements of life on the earth and in all of the universe. So that we then encounter ritually the real presence in all of our world, and especially in and through the gathered community, the real presence of the Spirit of Jesus, the Word of God we believe incarnate in Jesus, by which all things are made and constantly re-made by the eternal and unchanging God. The grace of this encounter, of this real presence to which all are unconditionally invited by Jesus, is itself forgiveness for the evil that we do, as it is also the empowering of us to reverse evil-doing and revert to good, generous living. Looked at so, the Eucharist is all of a piece with the self-giving, co-creative life that is especially characteristic of creatures made in God's creative image. In short, it is all of a piece with what we call morality, good moral living; it is the empowering centre of this. So that the Christian life to be lived by all the people does not fall into the separate categories of good moral living, on the one hand, and regular attendance at the sacraments, on the other. And we are saved the puzzlement that might otherwise have us wondering why in the judgement scene of Matthew's gospel, Jesus never mentions those who went or didn't go to Mass, but only those who did or did not live their common responsibility for each other.

Then too, the suggestion that the laity should take up fully their inalienable responsibility for advancing moral value in society in both its practical-living and its more theoretical-codifying forms, might lead quite naturally to the earlier suggestion that we should take now some initiative in inviting members of other Christian churches to our Eucharistic tables, and accept similar

invitations to theirs. Indeed, it could eventually lead to a growing suspicion that we need not confine invitations to the table of
the Lord even to baptised, card-carrying members of Christian
churches, much less to the sinless amongst these. Like Jesus himself when he instituted his table-fellowship during his public
mission – and not just at the last of his suppers on earth – we
should open the table of the Lord, as he did, to everyone, even
those considered the outcasts and evil-doers of society, as an
invitation to all to experience the real presence of that Creator
Spirit that we say incarnates in Jesus – an experience that has the
power to convert any and all to a life of self-giving love.

But I must end with the hope that, without this more radical
theology, or despite the somewhat convoluted considerations in
which I have sought to summarise it, a glimpse may still have
emerged, of a vision of 'Jesus Christ, Alive in his Church, Source
of Hope for Europe,' to take again the title of the Synod of
Bishops' working paper, through this brief effort to show in particular how the laity, appealing to nothing more than their
birthright, do daily engage in 'Serving the Gospel of Hope.' And
showing also that they can do this more effectively the more
they understand their inalienable responsibility to take their
own initiatives in promoting that moral living in which being a
Christian primarily consists. At the very least, this is a vision of a
church in which the church does not exist for the church, but for
the world and for the promotion of the rule of God, a rule of creative love, that is always already operative in the whole world.
And so, contrary to what the synodal working paper suggests,
the people of God, as they call themselves and who, as they say,
are the church, do not act first and foremost to build up the community of Christians in the world, much less their own frag-
mented version of that community, with the addendum that the
wider human community may afterwards somehow benefit
from this. If we did nothing more than reverse those priorities,
our vision of the future of church and world would have seen a
worthwhile improvement.

Concluding Unscientific Postscript

Scientists who engage in the debate with religion these days – although not all of them do this – commonly exhibit certain forms of conceptual confusion, accompanied occasionally by some logical ineptitude. The first exhibit can be called The Ulterior Motive Syndrome; the second, The Cheshire Cat's Grin Syndrome; the third, the Quantity for Quality Syndrome; and the fourth, The Duck-Rabbit/Matter-Mind Syndrome, or at least a diseased form of the same.

Before addressing the first exhibit it may be worth clarifying some terms that prove to be key terms, and constantly in use here. Basic to this terminology is Aristotle's definition of a 'thing', or rather of a 'nature' common to a whole species or genus of things: a form (of matter or of reality) that has within itself the source of its change or development (into successive forms). In talking about the universe we live in, this definition has the advantage of highlighting the activity that characterises all of creation and that is evident to us in the evolutionary coming-to-be of all that then makes up this universe. It also leads naturally to the description of the activity of a divine creator, if such there be, as a matter of forming, rather than shouting the universe into existence. And so Aristotle's definition has the general advantage of being able to accommodate the continual transformation (metamorphosis) that in modern times is increasingly understood under the heading of 'evolution'. So that continual creation-as-transformation and evolution can quickly be seen to be one and the same cosmogenic event. And finally, this definition allows for the use of the term, formulae, instead of the more metaphorical 'laws of physics', as a class name that comprehends and denotes, *ensemble*, all of those equations, measurements of cosmological constants, models of physical processes, and theories – ranging from theories for local areas and phases of the material universe to 'theories of everything'.

With that last class of theory ranging from an original effort to harmonise existing or future theories of the four fundamental forces of nature – gravitational, electromagnetic, weak and strong nuclear forces – to, as Deutsch for instance uses the phrase, a theory of a much higher, more ontologically comprehensive status that would account for the origin of the whole universe (multiverse or megaverse) and of all the entities-in-process of which it is constituted.

Formula then is the chosen term for any or all of these, whether individually or in inter-linked groupings. Because in this universe we always deal with entities-in-process or process-entities; and just as the knowledge of the form of a thing (its properties) can tell us much about the processes in which it engages, knowledge of its activities can tell us much of what kind or form of thing it is. Yet this use of the term, formula, as a class name for equations, models and so on, must not be thought to imply that a 'theory of everything' in its most expansive mode will prove to be merely a summary of all the formulae that apply in the different phases of matter, the different areas of science, and what are called the different levels of the natural universe, pre-biological, biological, and so on – a kind of super formula of which all others are the varied derivatives. For, as must shortly be observed with respect with those different levels (in one manifestation, different levels of complexity), from the most fundamental sub-atomic level to the highest form of life (this bottom-up imagery is metaphorical, of course), the formulae that describe and explain what goes on at a higher level are found to differ from those that obtained at a lower, even contiguous level. Even though the formulae operative at the lower level still remain operative in effect for that level as its presence within the higher level continues; for what we observe really is evolution, and not a series of simple replacements.

Formulae then, it must finally be observed, are entities of quintessentially mental provenance, containing imagery (strings, vibrations, bubbles, fields, and so on) together with abstract formulations (usually mathematical), as these exist in the creative agencies that are both involved in the continuous creation and engaged in investigating it. And as these describe and explain the entities involved and the processes by which these inform and transform themselves and each other as they adapt within the continual transformation of the universe as a whole.

The Ulterior Motive Syndrome

Scientists often present with the symptoms that are constitutive of this syndrome shortly after they become infected with the question of the purposefulness or purposelessness of the universe. Now this question is very frequently linked to the issue of emergence, with the suggestion that, if purpose can be found in this universe at all, it occurs only with the emergence of life, that is, at the biological level – if not only at a certain level of life, the human level. (Karl Popper once declared that the universe has no meaning or purpose, but we can give it one.) 'Emergence,' according to Philip Clayton, 'means nothing more than "that which is produced by a combination of causes, but cannot be regarded as the sum of their individual effects".'[1] Another way of saying that true novelties continually come to be in the universe and that, although when a novelty does evolve, one can look back and see the unbroken and apparently seamless connection of the steps that constituted that process, one simply could not have predicted the particular form of the novelty that eventually arrived on the scene. Clayton counts amongst the factors that, in recent science, favour this emergent view of our evolving universe: Heisenberg's uncertainty principle; the Copenhagen theorists' insistence that quantum indeterminacy reflected, not a problem of ours in seeking to know our world, but rather an inherent feature of the physical world itself; and chaos theory, which suggests that in a universe of exponentially increasing complexity, slight alterations at some of the more fundamental levels can cause quite disproportionate disorganisation at much more complex levels, causing disequilibrium and consequently requiring more novel and unexpected solutions in the form of adaptive self-organisation formulae at the higher level.

At a Templeton Symposium on Science and Theological Imagination, Simon Conway Morris commented on this phenomenon in evolution by talking of convergence. By convergence he meant 'that notwithstanding the riot of life on earth, its endless fecundity and seemingly limitless diversity, in point of fact old themes (or formulaic structurings) are played out again and again. Not only that, but complexity and novelty seldom emerge from brand-new structures, especially invented for the occasion. Much more usually, life seizes upon what is already available,

---

1. *Science and Theology News*, October 2004.

co-opting and re-engineering the existing "building blocks," albeit in ways that even to us seem surprising and unpredictable. The end-product, therefore, is a paradox because its process of emergence implies blatant jury-rigging and improvisation yet at the same time it shows, at least to our eyes, a seamless and beautiful adaptation.' One would simply want to add that life itself, biological systems as such emerge because, as Clayton puts the matter: 'The phenomena of living organisms (themselves) result from inter-connecting biochemical processes within very complex (dynamic, pre-biological) structures.' Convergence then is the other side of the coin of emergence: as different new forms of process-entities converge on the same existent formula, each in order to adapt more successfully by producing its own novel form of itself and its processes for survival. One of the examples given was that of the convergence of a number of different novel eye-like entities, all tracing back to a type of protein known as crystallins.[2]

It is at the level of the emergence of life that Clayton begins to talk about purposive activity and purposes in the universe. Why? Because, as he puts it, these biological systems 'base their future behaviour on their recorded memory of the past.' And what is their purpose in so doing? The basic example of such purpose given by Clayton, with a significance that we must shortly analyse, is this: 'surviving and producing the maximum number of viable offspring.' In view of the novelty that is a constant feature of emergence-evolution it might be better to express that same purpose in some such terms as: surviving and producing ever higher and more lasting forms of existence and life.'

Now the puzzling question concerning Clayton's account of emergence and purpose so far is this: why does purposefulness

2. Michael O'Keeffe describes emergence as follows: 'Although the higher level properties can be retrospectively shown to be consistent with the laws and principles operating at the more fundamental level, the reverse process – the prediction of emergent behaviour – is not usually possible.' Although he adds, 'retrospective computer-based numerical simulations of simplified models are possible today.' As Philip Clayton also notes: 'Computer simulations such as John Conway's "Game of Life" work with simple algorithms not unlike basic physical laws. Yet they can produce highly complex automata that move, interact and even reproduce as the game progresses.' *Science and Theology News*, October 2004, p 30.

appear only at the biological stage, when life forms have already come into being? Life itself may indeed be the greatest novelty so far produced in this universe, but for all that it is no less produced by the self-same kind of formulae operative at the pre-biological level as operate at the biological level: that is to say, by formulae that are sufficiently flexible or indeterministic to be capable of yielding or morphing into formulae for novel solutions congenial to ever-changing adaptations. So it surely can as rightly be said of the former as of the latter that 'they base their future behaviour on their recorded memory of the past.'

For the recorded memory in both cases is nothing other than these formulae, these dynamic processes which both entities or systems, the living and the pre-living, in a very real sense are. And if that is the case, then it can also be said equally of the pre-living as of the living systems, that they all have the purpose of bringing about the novel forms of the common stuff of the universe that they all share: the purpose, namely, of survival (in these novel forms also), of multiplying their viable offspring to the maximum, and of reaching thereby higher and more lasting forms of existence and life. In short, and especially where the purpose in question is described simply and satisfactorily as the furthering of the existence and the life of all that exists and all that lives, there simply is no valid reason for confining the attribution of purposefulness, any more than for confining the attribution of emergence, to the biological rather than the pre-biological stage of cosmogenesis. It is utterly illogical to even attempt to do so.

Furthermore, the whole universe as such must be said to be both emergent and purposive in the manner just now envisaged. And not just the separate stages, phases or levels of the universe. Or, even less so, the more specific *genera* or species of process-entities of which the whole universe is made up. Or, even still less so, the gene to which Dawkins in an atomist-reductionist view of biology that is as relevant to contemporary science as are the dinosaurs, seems to want to confine that purposefulness in the universe that finds expression on almost every page of his published work. The logical reason why both emergence and purpose must be thought to characterise the whole universe as such and not just the stages, levels, species or *genera* of which it is constituted is this: any formula that models any entity-in-process is seldom, if ever, confined to act enclosed in and for only one organ or organism, genus or species and, so confined, to bring about survival for that, and its novel emergence into the future. Quite to the contrary, the innate or infused formula that drives any particular process-entity is in touch with the surrounding environment, reciprocally 'sensing' it, gaining 'information' from it, and affecting it in turn as it directs the process-entity towards further and eventually novel existence or life.

The single-celled organism, *Rhodobacter sphaeroides*, 'senses' its environment and so moves towards the most likely sources of food. It does this through deployment of its proteins according to a process that can be modelled in mathematical equations. For proteins are the key workers in this enterprise. They 'communicate' with the outside world, transport things in and out of the cell, segregate the DNA between dividing cells, and so on. So from this simple instance alone it becomes clear that the formula that drives this particular process-entity, this single cell microbe, achieves its purpose of furthering existence and eventually evolving further forms of it, by operating in closest interaction with the complex environment of innumerable other process-entities.

What all of this amounts to and what corresponds with our common, broad experience then is this: that this universe is not a loose conglomerate of areas, phases, levels, *genera*, species, and individuals each working for survival and a novel future solely for itself and on its own terms. Such an entity, if it did manage to exist for any length of time, could scarcely be called a universe. No, all of these process-entities throughout the whole continuing process of emergence or cosmogenesis, are intimately interrelated and interdependent. Equivalently, the common purpose for which all process-entities operate must also yield the purpose of the universe as such, namely, the furtherance of existence and life without limit. For any species that is capable of refusal, to refuse to share in that universal purpose, and to substitute instead a particular purpose of enhanced existence and life for itself that entails deliberate destruction of other species involved, as our own increasing ecological disaster warns us, is a contravention of that species' own declared purpose. Equivalently also, the oft-referred-to meaning of existence and life in and of the universe coincides with its purpose. The advancement of existence and life in and of the universe through whatever forms it is due to pass, is what itself gives meaning and value to the universe and to all that is part of it. Or, in the case of conscious beings who are capable of this, the meaning can be described as the enjoyment of such existence and life.

But in either case no ulterior *raison d'être* is necessary, and no ulterior motive for those who need conscious motivation in order to engage with all other agencies in creation in co-operating for the promotion of existence and life. Indeed, once again, ulterior purpose and motive at this level are as like as not to promote destruction by those who entertain such ulterior motives, for they will inevitably release into the world a selfishness more destructive than the natural and wholesome selfishness that is held in harmony with the altruism that stems from the realisation of the interdependence of all in the promotion of existence and life for all. An altruism that even Dawkins shows to be exhibited by and compatible with his ever-so-clever little genes. An ulterior motive refers to an outcome of one's activity that is other than, additional to that which is achieved by that activity itself. So for example, if as is now evident from all forms of existence in the universe and from the formulae for these, their agency is designed to make them be and live more lastingly and fully together, then any motive over and above that, such as living for God's sole purpose in creating, to give glory to God for all eternity, would have to be considered ulterior.

It is at this point that the issue of a divine creator of the world arises, if only because in the relevant literature that issue is commonly connected with the issue of purpose in the universe. Philip Clayton formulates the connection as follows. If we were dealing with the pre-biological universe, he maintains, there is then as yet no real evidence of a divine creator. 'If there is a God, God might have designed the universe with complexity theory *in mind* (the theory of emergence, that is to say; emphasis mine).' But, he goes on to insist that nothing so far in what he calls the minimal emergence that occurs in the pre-biological universe could 'challenge a thoroughgoing physicalism and determinism, a universe without meaning or significance.'[3]

Precisely what scenario is envisaged here? Is it that what Clayton calls the minimalist emergence that takes place at the pre-biological level is so minimal that that universe is virtually the same as the classical physics of the 19th century would have described it: in the grip of a determinism so complete that all of

3. Clayton, loc. cit.

its future behaviour could be predicted from an adequate knowledge of the positions, velocities and so on of any point in the past? A universe from which, even if a god might be postulated in order to set the universe there and in motion by embedding 'the laws of physics' in it, the god could thereafter be excused from any presence to or in it, for it would simply run on its own? But it has been argued that, both before and after the emergence of life, the formulae for emergence are equally necessary and of the same kind, substance and significance. If only because life too emerges in the natural way in which all emergence takes place, and requires no special introduction and no more sense of purpose and meaning than anything else that emerges before it or within its realm.

Furthermore, in both the pre-biological and the biological stages it is equally the case that, in the course of any particular emergence, the formulae previously in force do, to some extent, determine the newer formulae that will emerge as the descriptions and explanations of the novel entity-processes. For these can be seen, if only in retrospect, to have positively led to the new entity-processes, and at some level of these the former still apply. But yet these formulae that precede a point of emergence simply cannot be said to determine absolutely what will emerge, as could happen in classical physics. Therefore Clayton's picture of a universe, as yet innocent of biological processes, responding well to Laplace's reported remark when asked about a god: 'I do not need that hypothesis', is simply not acceptable on Clayton's own terms. For even if the existence of such absolutely deterministic formulae as we see at work in classical physics did not require the presence of a creator god in the course of the history of that universe, at least up to the point at which life somehow emerged, and the universe only at that point ceased to reveal 'a thoroughgoing physicalism and determinism' that characterised its previous phase – and that is highly debatable – it is certainly the case that formulae which are only conditionally or partly determinative of what will emerge would seem to require the constant presence of an applying mind-like entity – precisely for the continually creative rather then absolutely deterministic functioning of these formulae and of the forms they help fashion in ever novel ways.

Therefore no stage or level of the universe represents a thoroughgoing physicalism (that is, without any element of the presence of mind-like entity, as distinct from mere mindless matter), and consequently a thoroughgoing (complete or absolute) determinism. For the same kind of inherently creative formulae govern the universe at the levels from which life emerges, and at the biological level. Formulae / forms affected by a certain amount of indeterminacy / chaos are, when one thinks of it, necessary to creativity, which at any level connotes the novel and the new. As Nietzsche once put it: one must have chaos in order to give birth to a dancing star. Such formulae / forms are necessary at any level, from what physicists used to call the most fundamental level at the sub-atomic phase, to the most complex level represented in this universe by intelligent life. And that, incidentally, is the end of the hard reductionism of late classical physics, however it may still be pursued by hardies like Dawkins and his very clever, if also very selfish little genes.

Now the issue of divine creation seems to have arrived, with these references to a mind-like entity whose presence seems necessary at all levels and stages of the existing universe. But before turning directly to the god-question, in connection with another syndrome that affects it, some concluding thoughts on purpose are called for. First, then, the purpose that is revealed in the actual conduct of all natural agencies involved in the created realm as we know it, appears universally to consist in the motive to survive by an adaptive transformation that in turn results in life, and indeed in ever more complex or higher forms and experiences of existence and life. That then, it would seem to follow, would also be revealed as the fundamental, non-ulterior purpose and motive of a divine creator creative agency also, if only because that agency appears to work always in and through the whole universe – in and through all of its then derivative and dependent agencies.

And yet, if any fact of the matter is more obvious than that just stated, it is this: the protagonists on both sides of the current science-religion divide quite frequently assume on the part of the divine creator, whom the one side denies and the other as ardently confesses, a motive that is distinctly ulterior indeed. Those from the science side may say that they pick up these divine ulterior motives from the other side. And that is fair enough, for the other side do rather go on about their privileged authority on such matters. And undoubtedly the most common version of the ulterior motive that, according to Christian preachers, persuaded God to create this world, was his own eternal glory, a glory that needed constant and eternal acknowledgment by the human race. More graphically, so that we humans, who can detect and appreciate the wonders of this world, should respond by singing God's praises for all eternity. For a picture of this particular divine ulterior motive for creation, but now used from the science side and with the ulterior motive on that side of painting God the Creator as a sadistic, self-obsessed megalomaniac, simply look back to Bertrand Russell's quotation from Mephistophelis' speech, in Chapter III above.

What is to be said from the religion side about this alleged ulteriority of divine motive and purpose in creation? Simply, on behalf of natural philosophy as science informs that discipline, and as science ushers our inquiries at the very least to the borders of that natural philosophy at which the adjective, theological, begins to apply, that what is revealed in evolution/continuous creation is a purpose and motive of all agencies involved that is adequately represented as the natural outcome of all of that agency itself, namely, the limitless creation of existence and life ever more fulfilled and lasting. It is indeed possible to argue from the Christian part of the religious side (and I think from many, if not all other religions also, as I tried to do in *Christianity and Creation*) that the faith of the historical Jesus, being a creation faith through and through, reveals also a God whose only purpose and motive in creation is to thereby confer life and life ever more abundant on all that comes to be. A God of grace, then, for grace means a gift (of existence and life) freely given and for eternity. That is the portrait of divinity and divine motive revealed in creation, rather than the portrait of a sadistic divinity bent on eternal punishment for whatever sins we poor mortal creatures might manage to commit, or a God intent, first and foremost, on self-glorification. Although, naturally, all who have been graced would find it part of their moral response to be grateful to the divine benefactor, as a first step and motivation towards taking up their own responsibility to promote existence and life for all.

Turn then to what is known as 'the god question,' for the simple reason that god (*theos*) or divine (*theios*) is simply the name for whatever is discovered to be the ultimate of source of the universe, and ultimately responsible for it: for its origin and for those formulae that continue to guide its continual evolution, its continuing coming-to-be. Now the dominant assumption from early efforts by the first Greek physicist-philosophers to set out on the scientific venture of cosmic discovery, was that the ultimate source of this ever-changing universe was of the order of mind. And that dominant assumption seems to be shared to this day, even by those who so often in the name of science try to dismiss theology at the very outset from the range of investigation and discovery within which science in modern times has made such dramatic progress. And it is as we turn our critical minds to this kind of dismissal that we soon sense the presence of the symptoms of yet another logical disease rampant amongst this set of scientists: The Cheshire Cat's Grin Syndrome.

The Cheshire Cat's Grin Syndrome
In one of those popular books written apparently for children to describe some wonderland, but which in reality reveals to adults also some salient features of our common earthly lives, there is a scene which depicts a Cheshire cat up a tree with a grin on its face; then the cat disappears, but the grin remains in the same now catless branches. A metaphor here for the manner in which so many scientists seem to talk about the 'laws' of the universe. The word 'laws' is in parenthesis here as a reminder that it is a metaphor. For what is referred to in actual fact is that close-knit family of formulae that seem to guide the continuing coming-to-be of the universe towards a future that seems nevertheless to have been to some serious extent open in the past, and to show signs of continuing to be open towards its particular future.

Now the kind of talk indulged in so frequently by scientists in their excursions into deep cosmology, and which makes our Cheshire Cat's Grin Syndrome so apposite is this: the laws of physics that bring about the evolving universe we continue to investigate are still not so completely within our grasp that we can account for its coming-to-be up to this point, never mind for the future. Although we are now able to understand the dynamic nature of these laws, on the model of formulae that can bring about real novelties in the complex environment in which they are effective. Now many, many statements about these laws convey and confirm the impression that they are some sort of stand-alone agencies responsible for the coming-to-be of our still evolving universe – stand-alone like the grin of the Cheshire cat. But these laws of nature are simply not such stand-alone agencies, and the conceptual confusion involved in presenting them as such, together with the illogical conclusions then drawn from it – mainly conclusions congruent with a radically materialist and generally atheistic kind of philosophical position – need clarifying, to say the least. And the best way to begin to clarify this confusion of concept and logic is to observe that these laws, or rather the formulae of which they are made up, are of the nature of mind. That is to say, they exist only as functions and progeny of mind; so that, whenever and wherever they are seen to operate, mind is operative also.

If one gave little enough thought to this matter, one might be inclined to think that one of these formulae could be said to exist on this page as soon as there appeared on this page the following equation: $E=MC^2$. But a very little more thought would suffice to persuade any reasonably intelligent person that this is not the case. For these print-marks on this page are not in themselves the formula. Well, no, one could agree, but they are the expression of the formula – a set of signs, in this case made up of letters and numbers, that signify the law. So, the formula is there on the page in that sense? Not really, for the simple reason that these marks on the page, $E=MC^2$, can function as signs only when they exist, not on the page, but in the inter-subjective commerce of at least two minds of the kind that created these signs in the first place. For signs exist only as functions and progeny of mind. Lying on a page, unread by a mind wishing to signify something and unread by another mind that agreed to share the convention of just this sign, these characters drawn on a page are no more than a meaningless squiggle.

One could of course adopt the metaphor of writing as inscription, in order to say that the equation, $E=MC^2$, is inscribed into the corresponding process of the constant alternations of mass and energy in the mass-energy complex that is our universe. But then one dare not forget that this is a metaphor, and that a metaphor consists in collating two quite different images which yet have something in common, in order by such comparison-contrast to illustrate further what is under investigation. So in the present case, after very long investigation of the apparently infinite complexity of the natural world, some greatly insightful person sees that one deep part of that process-reality is governed by and so reveals the elegant, indeed beautiful simplicity of a formula, intelligible to mind because it is itself an issue of mind, expressible in five common and conventional signs.

The metaphor then says that this equation was read from or in the evolving universe, as if that were a book from which one had taken a leaf, and that Einstein had 'read' it there, so that some later reader could read it on a page from a paper that Einstein had written. A striking metaphor for the more literally expressed fact that the mind of Einstein had come to know and understand a process in nature that a greater mind than his formed after the manner expressed in this equation. Therefore even in this metaphorical collocation of the quite literally understood imagery of reading signs on a page, with the imagery of Einstein's investigation of the fabric of reality, it still remains a matter of mind understanding what mind has made, when it is communicated to another mind. These equations or laws of nature are functions and issues of mind, expressed in the first case above through mind's choice of conventional signs, and in the second case by mind's fashioning of a particular process according to the equation or equations at issue.

At this point of the clearing out of conceptual confusion and logical ineptitude, then, can it be concluded that some mind-like entity creates the universe, and by doing so earns the title of divinity? Not quite yet, must be the answer to this question. Because another syndrome can be detected, symptomatic of a sickly bout of reasoning that is, like Dawkins' desperate atheism, a throwback to a past era of scientific philosophising known as the Age of Heroic Materialism. This is The Quantity for Quality Syndrome, the key symptoms of which consist in the confusion of thinking that an increase in the quantity of purely quantifiable entities can of itself account for an emergent *quality* such as consciousness, for example, with its emergent and characteristic capacities for emotion, imagination, thought and free will.

The Quantity for Quality Syndrome
David Deutsch's otherwise impressive work, *The Fabric of Reality*,[4] provides a striking instance of just this syndrome, and does so coincidentally in his account and defence of the freedom of the will as an obvious emergent feature of our known universe. According to Deutsch, to have free will means 'that we are sometimes in a position to effect future events (such as the motion of our own bodies) in any one of several possible ways, and to choose which shall occur; whereas, in contrast, we are never in a

---

4. David Deutsch, *The Fabric of Reality*, esp pp 269, 338.

position to effect the past at all.' He then makes two accurate and correct observations about this freedom of the will: first, that 'freedom has nothing to do with randomness', and, second, that 'what we think of as our free actions are not those that are random or undetermined but those that are largely determined by who we are, and what we think, and what is at issue.' For 'we value our free will as the ability to express, in our actions, who we as individuals are.'

But then on his way to his final explanation and defence of free will in our universe, he makes a number of other observations that are rather more questionable. He claims that 'in any classical world picture ... a typical statement referring to free will... "I chose to do X; I could have chosen otherwise"... is pure gibberish.' For he represents classical spacetime as a 'static block universe' in which time flows evenly from the past into the future, and so future events are 'already there' in that picture of the universe. It is not altogether clear what that means, and the confusion is compounded when he declares that 'the difficulty of reconciling free will with physics is often attributed to determinism, but it is not determinism that is at fault. It is ... classical spacetime.'

In our common understanding of the classical world picture painted by modern science, it was indeed the determinism of the laws of the physical universe that seemed to rule out free will. For the future did not already exist for us, and if God who by nature transcends time knows everything in God's eternal present (represented perhaps by the drawing of a static block universe?), even what to us is future, that divine knowledge does not as such cause the events. God causes particular events in creation through creatures who cause them. Therefore the free will of a creature that can act freely is preserved. Furthermore, in the block universe of classical physics, there are real differences between points of time and spots of space, in that it is possible to travel from one spot in space to another while going from one point in time to another. And all of these do not collapse or coalesce into one point-spot. Correspondingly, eternity does not refer to a limitless extent of such spots and points, but to the *'simul ac tota possessio'*, the simultaneous and total possession of, or presence to all, that is the prerogative of the timeless and extensionless source of all. For us finite ones who live within space-time, even in a block universe, the future is really not 'already there', any more than it is 'already here'. So that it is the absolute determinism of classical physics that challenges the existence of free will and not, as Deutsch wants to maintain, the block universe as such.

This piece of confusion is then followed by Deutsch's own explanation and defence of free will according to the quantum physics that replaces the recently classical worldview. This explanation and defence depend entirely upon the multiverse hypothesis that Deutsch argues for and embraces. And it is here that sheer quantity takes over the burden of proof of the existence of the quality of freedom of the will. Briefly, 'I chose to do X' translates into 'Some copies of me in certain universes, including the one speaking, chose to do X.' 'I could have chosen otherwise' translates into 'Other copies of me in other universes chose otherwise.' He then goes on to offer other translations of standard statements that highlight the moral nature of free will choices: 'It was a right decision' translates as 'Representations of moral values that are reflected in my choice are repeated more widely (i.e. there are more of these) in the multiverse than those of rival values'; 'I am good at making such decisions' becomes 'Copies of me who choose X, and who choose rightly in other situations, greatly outnumber those who do not.' Quantity, not quality, now blatantly defines moral value.

Now that really and truly is gibberish; discourse saturated throughout with obtuseness, confusion and illogical argumentation, as a few salient examples from that argument must suffice to show. First, what precisely is meant by a copy of me in another universe? A clone? An identical twin? Then how did they get there? And even if it was explained to me how I was cloned, or how a twin of mine turned up in another universe, still as we know from clones and twins in this universe, they do not face the same sequence of situations in life, and with every difference in that respect, before taking into account at all the differences in the choices they make where they do face roughly similar situations, there is a decline in the degree to which they can be called copies of me. Perhaps it is holograms he has in mind? But then they would qualify even less for copies of me, for I am not a hologram. But hold now … The more one questions nonsense like this, the more one realises that one is showing it more respect than it could ever deserve. Whatever about the multiverse hypothesis itself, the idea of a multitude of universes (all or most or very, very many) containing copies of what is in others must surely be more a piece of science fiction than a credible piece of science.

If Deutsch has not explained and defended, or does not simply know from experience that choices he makes in this universe are free in the sense defined, then he cannot know or even hazard a guess that choices made by the inhabitants of other universes that are copies of himself in respect of their having chosen X as he did, might also be copies of himself in that they chose X freely. And this for the very simple reason that it is within the confines of this universe, it is from my experience here of the freedom I have in choices that I make, that I infer the freedom of fellow humans in the choices they make. It is as obvious as the similarly shaped noses on the faces of Deutsch and of all copies of Deutsch in all other universes that have copies of him, that it is because he experiences such freedom as he has in his own choices that he can have any confidence in the hypothesis that his many copies in other universes enjoy a similar freedom. The pretence that he knows of and can explain and defend the freedom of his will from the activities of his supposed copies in supposed multiples of other universes is just as preposterous as it sounds when read out loud.

Now this particular excursus into the manner in which a purely quantitative process of enumerating and multiplying instances is thought to bring about quality in conceptual analysis and logical argument, with all of its resultant nonsense, is not in itself relevant to the logic of this chapter and of this collection, namely, the logic of divine creation as the deep explanation of cosmogenesis. It is recorded here simply to show that the critique of certain explanations of the origin of the universe which, like freedom of the will in Deutsch's book, depend upon a simple multiplication of universes to infinity, or as close as dammit to infinity, is not being attacked by means of a kind of critique especially invented for the dispute about creation alone.

Turn then to the case of the formation of universe(s). For quite some time now, there has been talk of the so-called anthropic principle. Roughly speaking, this means that for all intents and purposes the only universe we directly observe seems to have been fine-tuned to the precise end that life, and more specifically life in some such form as that of *homo sapiens*, should emerge to observe it. That is to say, that the so-called laws of physics, and even the mathematical values of certain so-called cosmic constants, would have to be what they are, or very close to that, for this universe to come about as it did. Susskind takes the example of Einstein's cosmic constant, a certain universal energy that acts on all matter. If its number or size were different neither life nor ourselves would be here; and if it were somehow altered, both would disappear along with much else that characterises the current universe. So Susskind freely allows that 'the appearance of intelligent design is undeniable.' Yet we are quickly disillusioned if we think that 'appearance' in that allowance means 'arriving on the scene'. Quite to the contrary, as the title of Susskind's book makes initially and permanently clear: *The Cosmic Landscape: String Theory and the Illusion of Intelligent Design*.[5] 'Appearance' means 'illusion'. In other words, all of that fine tuning was designed (that word does keep intruding itself, even where it is obviously not wanted), not only to bring us into existence, but to subject us to what must surely be the grandest of all illusions – an illusion, quite literally, on a cosmic scale.

Dr Susskind's prescription for ridding us of this illusion, induced in us by our unkindly universe (or, heaven forbid, by some darker, more malicious and hence personal power to whom the universe and ourselves are mere puppets), is in two bottles. The first is labelled String Theory. According to this theory the most basic and universal building blocks are not particles but one-dimensional 'strings'. These strings 'vibrate' apparently in so many different 'modes', and these modes 'represent' all of the different particles that come and go in the formation of the universe. And that accounts for the 'landscape', the cosmic landscape of Susskind's title. (All of the foregoing parentheses are designed to draw the reader's attention to the multitude of metaphorical or, since these are part of a cosmic story, mythic images now as ever in use in 'hard' science.) This is a landscape consisting of all possible environments, and that apparently includes all possible universes. And thus and so to the multiverse or, as Susskind would prefer to say, the megaverse.

But 'Halt!' cries one reviewer of Susskind's book, Michael Duff, at this point, 'in order to avoid confusion', as if there could be no possible confusion in the matter up to this point. 'The landscape', warns Duff, ' is not a real place, just a list of possibilities', presumably all the possible environments referred to already above, that are 'permitted by theory' (that is to say, rationally understandable). So what we have then is 'a landscape of possibilities populated by a megaverse of actualities.' And now we have opened for us the second bottle of Susskind's prescription to be taken regularly against the onset of chronic states of potentially religious illusion. There is a multiverse after all – is this simply because string theory envisages such a possibility, or did we miss some evidence along the way of this argument that would persuade us a little more positively that more than one of these possible universe-like environments has been in fact made real? Only if we leave off these niggling interruptions will we be rewarded with an argument from the fact of a multiverse, an argument different in kind from the argument which Deutsch produced in his attempted defence of freedom of the will, but similar in that the very existence of a multiverse explains something which might not be explicable in the case of a uni-universe, other than by a potentially theological form of explanation, real intelligent design.

Here now is the explanation of this universe which gives the illusion that it was intelligently and especially designed so that we should emerge; encapsulated by Duff in a quote from Susskind, and followed by a comment of his own. Susskind is explaining the 'how come?' of this universe by concentrating on the 'how come?' of this cosmic constant with just this value, without which we would not be here. Question: 'Why is a certain constant of nature one number rather than another?' Answer: 'Somewhere in the megaverse the constant equals this number (possible); somewhere else it is that number (real). We live in one tiny pocket where the value of the constant is consistent with our kind of life. That's it! That's all. There is no other answer to the question.' And Duff adds: 'The anthropic principle is thus rendered respectable and intelligent design is just an illusion.' To which one wag added; thank God for that! God is gone and we have our atheistic or at least agnostic respectability back, if not much else.

In an age in which scientists are credited with a near monopoly of all the truth that is available to us mere mortals, and in which their pullulating successes come near enough to a full authorisation of such credit, it might seem foolhardy to query Susskind and others who query intelligent design while going about their normal scientific business. And particularly fool-hardy to query an explanation that ends in a peremptory, 'That's it! That's all.' So before doing just that it might be well to ac-knowledge that in some of the matters presently involved Susskind and his kind are unfairly criticised. They are criticised, for example, for surreptitiously leaving the domain of science at points where ordinary scientific verification of the straightfor-ward kind obtains, and drifting, if not at times quick-marching, into areas well beyond any real and present prospect of verific-ation. Areas inhabited by 'strings', 'pockets', 'bubbles', 'attrac-tion', 'energy', 'megaverses' and so on. They have left the realm of science for the realms of cosmic myth. Or, when they use the language of conceptual abstractions, like mathematics, instead of using the imaginative imagery of strings and so on, they might be said to be headed for a realm beyond physics that is aptly called meta-physics.

But is it fair to be critical of a move from physics to meta-physics? From the very beginnings of Western philosophy meta-physics followed in seamless continuity from the early physics, the study of the *physeis*, the natures of all things in the natural world. Metaphysics followed quite naturally as the quest was pursued, beyond the observation of particular laws that matter in motion in the natural species allowed to be seen, and towards the deeper and more universal features and factors in reality that were not quite as directly observed, but were thought to be necessary in order for us to understand how the more observ-able universe of natures came to be what it is. On some such un-derstanding of the matter, that metaphysics be descriptive rather than prescriptive, some advocates of the British empiricist tradition of philosophy have regained a sufficient degree of their native intelligence as to allow metaphysics once more into the philosophical fold. So that, as of yore, it can follow on prop-erly and seamlessly from the physical (and other) sciences. And who better to pursue this follow-on from the physical sciences than the physicists?

That having been said, however, it is necessary to comment on what appears to be a certain lack of conceptual clarity, to-gether with some logical *non sequiturs* in the kind of case argued by Susskind above. As a case of very possible conceptual confus-ion, take his landscape of 'possibilities' that are not (yet?) actu-alised. And ask, what kind of thing is a possibility? On what kind of landscape do these grow? Where did you last encounter one? In a mind, of course, is the answer. A possibility that is not (yet) a reality is born of a mind-like entity that has, and knows that it has creative possibilities in a universe that is not governed by laws of physics so pre-determining that nothing novel can be made in or of it. In this respect, possibilities are similar to the so-called laws of physics that govern the emerging universe(s). These too, as has been argued already, are mind-stuff. It is by such image- and idea-like entities that we define mind, and note instances of its existence. Minds are made up of the stuff of ideas, forms (as Plato would call them), formulae, equations with values attached, and so on. As such, as mind-generated entities they govern the emergence of the universe, and they can therefore be abstracted from it by minds that can act creatively with and within it. And how far are we now, in this respect of a designing intelligence type of creator, from the cosmologists of old who talked of world-souls fashioning the world from within, or of worlds created first in the form of ideas, in the form of designs, both formulaic designs and image designs in the mind of the creator, and then created in material reality (with the words 'first' and 'then' understood to be conditions of our human perception of the process, and not necessarily descriptive of any time-lapse in the originating one)? So that it is at this point of clarification of his concept of a landscape of possibilities that we can really agree to his 'That's it! That's all. There is no other answer.'

Yet Susskind and his kind would still like to drive us as far as possible from such an illusion, as they would call it. An illusion of intelligent design foisted upon our unwary minds, he must now think, by the very process of cosmogenesis he has so carefully described. But by now the only stick he can threaten to drive us with – let us mix metaphors here in quite a cavalier fashion – is in the second bottle of his prescription for the removal of illusions, the multiverse hypothesis. But this does not, and logically it cannot, work. For if it is the coming-to-be of our very own universe that makes it so that 'the appearance of intelligent design is undeniable', how could the simple multiplication even to infinity of universes presumably like ours, even if only to the extent that they are universes, decrease rather than increase the case for calling this an illusion? Until Susskind would have to concede that it was never an illusion at all? For each uni-

verse, if it is truly a universe, working as a unit through the ordered, 'law-abiding' co-operation of all its parts and elements,
and each governed by laws of physics, however locally different
these laws and values may turn out to be in each case from those
that obtain in our universe, would surely strengthen the veridic-
ality of the appearance of intelligent design, and then of
Intelligent Designer(s) (see the Cheshire Cat's Grin syndrome
above).

Someone might say at this point that the foregoing critique is
over elaborate, so that it simply missed the simpler claim made
by those who wish to show that intelligent design in this universe is an illusion. For these are simply arguing that, given a
sufficient number of universes coming into existence, the coming-to-be of this one, that is apparently designed to produce us,
is sufficiently explained on the grounds of sheer probability. But
that vastly simplified form of the argument would work only if
not just our universe, but all universes came about by random
combinations of whatever particles form the basic building
blocks of universes in the random presence of other random factors, each with randomly assigned values and perhaps randomly
altering values. Yet, because of the quite literal impossibility of
conceiving how this infinite plethora of particles and other factors could exist before coming together to form universes, or
how their original separateness could be envisaged in such a
way as not to prevent their ever coming together, except perhaps spasmodically and in small groups accidentally occurring,
it is not at all feasible to try to explain the formation of any universe, not to mention a multiverse, in this fashion. In short, either all of the other universes in the multiverse are true formula-
designed universes, and then in every single case 'the appearance of intelligent design is undeniable', and as the undeniables
increase exponentially in number, intelligent design approaches
the status of absolute undeniability, and 'apearance' then means
'coming into (our) presence'. Or else none of these are true universes, just accidental collocations of particles and so on; in
which case none of them can count towards numbers required
for statistical probability about universes and how they come
about.

Now that last remark would certainly seem to suggest that
scientists who advance this vastly over-simplified version of an
argument against design are still suffering the mental hangover

from an age of crude reductionist materialism. This has been traced back at times to the 'atomism' of the early Greek physicist-metaphysicians, Leucippus and Democritus, perhaps to give it the weight of antiquity and the prestige of a noble lineage. But even Leucippus and his followers in that Atomist school of ancient cosmology were very far from even suggesting that a random encounter of particles could ever account for any genus or species in the cosmos, not to mention accounting for the cosmos as such. Chance or randomness in the encounter of particles, or in the mutations of these, could not account even for *Rhodobacter sphaeroides*, with the diffusion of its proteins within and without in order to find and assimilate suitable food in its highly complex interaction with an almost incredibly and increasingly complex environment. So, although they never heard of Rs, and would not recognise one if they tripped over one in the corridor, Leucippus and his followers insisted that it was the shape or 'form' of the atoms that played the key role in their collocation that resulted in cosmos. What we would nowadays call the properties of particles and of the powers of attraction or repulsion, together with emergent formulae for adaptive interaction in an increasingly complex environment, that can be seen at least in retrospect, and that add up to our so-called laws of nature – that can be modelled mathematically. An intelligible entity then, a thing co-natural with mind, for mind and of mind. Leucippus in actual fact acknowledged this when he said that his atoms interacted, not by chance, but rather *ek logou kai 'up anankes*, as an issue of reason and on a basis of necessity. Aristotle said that 'Leucippus and Democritus in effect make all things number, and produce them from numbers', a dim premonition perhaps of the mathematical modelling that reveals the designer universe we now know so much better?

One might well suspect a similar piece of fraudulent logical sleight-of-hand in another version of cosmogenesis in contemporary physics, a version now which does not necessarily involve the still very highly speculative multiverse hypothesis. This version begins from a scenario in which the initial conditions of the universe (its original 'flatness,' its homogeneity throughout, its 'smoothness' and lack of internal defects) join the fundamental laws of physics both in sharing the property of 'transcendence' (that is to say, applying across all categories, kinds and levels in our emerging universe), and in the extraordi-

nary specificity which is required, at least in some of the constants involved, in order to give us the universe we now see.
And once again with the consequent and apparent entailment of
intelligent design.

To quote Michael O'Keeffe on the subject: 'Physicists were
not happy with this state of affairs. (Why? Because of the support it gave to intelligent design? The next sentence would seem
to suggest so). A way of dispensing with the extraordinary initial precision was sought, which would allow the development
of the universe, with the properties it is known to possess, from
a wide diversity of starting situations. The concept of inflation,
devised by Alan Guth and Henry Tye of Cornell University, met
all the challenges. The inflationary theory of the universe illustrates how the large-scale universe that is observed today could
develop from a small 'seed', part of a larger whole, whose initial
conditions did not have to be specified too tightly. Given the
small size, a reasonable initial uniformity could be expected;
and then if an early exponential expansion took place, today's
universe with its properties of flatness, smoothness, isotropy
and homogeneity, would naturally arise without any element of
mystery or surprise. According to the theory, the universe began
– without particles, without matter – as a zero-valued Higgs
field. This represented the presence of a very high energy density
in the universe. It was a highly unstable state of affairs. Very
early on the field acquired a non-zero value. This led to an extraordinary brief, but stupendous, accelerated expansion of the
universe, ending with the energy being converted into a spec-
trum of exotic particles and providing the 'bang' for the Big
Bang.'

The give-away phrases in this clear description of inflation
theory are: first, 'allow the development of the universe with the
properties it is known to possess, from a wide variety of starting
situations;' and, second, 'develop from a small seed, part of a
larger whole, whose initial conditions did not have to be specified too tightly.' Placed in such a context, it would almost seem
as if the cosmological theory of inflation was designed (intelligently, of course) to make way for much less specialisation in
the nature of the initial conditions for the universe, with a consequent increase in the number of possible starting formats for different universes that could then emerge. With the result that –
and here comes the desired conclusion – amongst so many pos-

sible universes emerging, the probability of our particular one emerging would be sufficient to avoid the otherwise near-necessity of attributing its emergence to the instance of a very specific intelligent design.

There is another issue that arises in this project of inflation theory that produces such numbers of possible and varying initial conditions as would make the emergence of our universe, especially as a flat universe, explicable on probability criteria, where one might otherwise have to invoke intelligent designs and designers. It is this: one would need to know a great deal more about the nature of matter and energy in order to be so sure of one's calculations for a flat, open or closed universe, as to arrive with any confidence at the preferred option of the flat universe. Yet what is striking is how little is known about matter or energy. On present calculations, visible matter makes up only 0.01 of the stuff of the universe (I have been calling this 'gross matter,' the kind of which our bodies are made); 'dim' ordinary matter, 0.05; and 'exotic' dark matter (its nature unknown, apart from its being an hypostasised explanation for the movement of stars), 0.36. To this may be added 'dark' energy which, if it amounts to 0.64 of the stuff of the universe, may prove that we have a flat universe indeed, expanding eternally but at an ever decreasing rate. However, with dark energy the case is even worse than it is with dark matter; for we do not quite know what energy is, never mind dark energy.[6] Do we even know enough

---

6. 'It is important to realise that in physics today, we have no knowledge of what energy is. We do have a picture that energy comes in little blobs of a certain amount. It is not that way. However there are formulas for calculating numerical quantity, and when we add it all together it gives ... always the same number. It is an abstract thing in that it does not tell us the mechanism or the *reasons* for the various formulas.' Richard Feynman, Leighton and Sands, *The Feynman Lectures in Physics*, Addison Wesley 1966, vol I, ch 4, p 2.

'Energy is a purely abstract quantity, introduced into physics as a useful model with which we can short-cut complex calculations. You cannot see or touch energy, yet the word is now so much part of daily conservation that people think of energy as a tangible entity with an existence of its own. In reality, energy is merely part of a set of mathematical relationships that connect together observations of mechanical processes in a simple way.' P. C. W. Davies and J. R. Brown (eds), *The Ghost in the Atom (a Discussion of the Mysteries of Quantum Physics)*, Cambridge University Press 1993, p 26.

to be sure that energy belongs to the category of matter rather than mind? Especially in a universe in which what we know and name as mind or mental stuff seems to be always embedded in matter, and vice versa. In Paul Davies' words, 'energy is merely part of a set of mathematical formulae,' themselves abstract mental constructs.

For the moment the uncertainty about these very entities in question here is merely meant to strengthen the adverse criticism of the argument from quantity already outlined above. For these criticisms exhibit the same strength when faced with inflation theory and the numbers of varying initial conditions for universe(s), as they exhibit when applied to a plain multiverse theory as such. It could still be true, if we could see these other possible universes that emerge from initial conditions different from our universe, that each would be as specific for its ultimate emerging outcome as is the case with our one observable universe. And it would in any case be true that a set of such intelligible initial conditions and fundamental physical laws, being mental constructs and hence intelligible in themselves and in their results, would entail the intelligent design of each universe in turn. So that the multiplication of such examples of cosmogenesis would increase rather than decrease the force of the argument for intelligent design.

So where does all of that leave us then? It leaves us all, hosts as we are to The Duck-Rabbit Syndrome, potential victims to a diseased form of that syndrome. The Duck-Rabbit Syndrome refers to a drawing by Wittgenstein which one can look at first and see it clearly as a drawing of a duck; then look at it again, or simply stay staring at, and see it suddenly and just as clearly as a drawing of a rabbit. The relevance of this to this present context derives from the fact that certain scientific cosmologists, ranging from early Atomists to modern physicists, and at least up to the point at which physics was still mostly particle physics, were self-styled, persistent and most dogmatic materialists, claiming to see nothing in the universe but matter-in-action (Russell's Omnipotent Matter rolling on its relentless way), even though they were also seeing the mathematical models always operative in nature. Whereas some other physicists, looking at the same universal phenomenon, claimed that they were looking at mind in action, their own minds in fact, acting apparently effectively on something they called the material universe, but of which they could not say that they had any really objective knowledge. More recently this very distinction itself between matter-energy and mind-spirit became more problematic when, as Patricia Williams put it, matter which used to be made up of 'small, indivisible, hard inert entities that interact deterministically,' then began to seem more like 'a form of energy that is ineffable, uncatchable, hazy, active and responsible for life', the very characteristics that once more or less defined spirit, soul or mind.[7] Or, more generally, this very distinction between matter and spirit, that is at the heart of the rejection of intelligent design from this unquestionably material universe, became questionable when physicists began to talk so confusingly about matter and energy.

The Diseased Form of the Duck/Matter-Rabbit/Mind Syndrome
The disease consists in the presence of the dogma asserting that only matter with all its emergences, and all of these equally and exclusively material, can be seen in the universe. A dogma, it must be said, that is just as dogmatic, resistant to reason and reactionary as any dogma ever entertained by the most self-proclaimed and exclusively god-inspired of religious fanatics. Just as if someone were to be dogmatic about Wittgenstein's drawing, to the effect that only a duck could be seen in it, and never under any circumstances a rabbit. This disease grew and survives only in a culture of hard dichotomous dualism, according to which matter and mind are two quite distinct kinds of stuff or substance, a dualist culture rampant in modern times and falsely attributed to Descartes.[8] For only against such a hard dualist background can our materialists maintain that what one sees, if it is matter, cannot also be mind, and can have nothing of mind in or about it. Despite the fact that our materialists, both modern and ancient, sometimes in what must be regarded as unguarded moments can be heard to sound as if indeed these intelligible designs by which the world comes to be are widely considered the very hallmarks of mind. And when queried about this they then usually (as in a recent BBC series on the brain) insist that when they use the common word 'mind', they are really referring only to the physical brain, and to nothing other than, or in addition to

---

7. *Science and Theology News*, October 2004, p 29.
8. See *The Critique of Theological Reason*, ch 1, section on Cartesian Philosophy Today.

that. Yet, for all that, the formulae referred to as the basis for belief in intelligent design are the very hallmarks of mind: they
come to mind, and come from mind; so much so that mind is
normally defined as the entity which both processes these formulae, and which gives origin to them in the first place. 'Mind'
is what we mean by an entity that processes such intelligible
formulae.

The question that now remains to be answered, therefore, is
this: what account of the relationships between mind and matter
in the universe could enable us to return to a healthy, rather than
a diseased form of The Duck-Rabbit Syndrome, in which we
could once more see in this universe, this object of our most constant contemplation, both mind and matter? (The thoroughgoing
idealist is just as diseased in this respect as the thoroughgoing
materialist.) The best answer to this question, it would seem, is
to be found in the image of reality as process. A process that
links in an unbroken chain ultimate source to ultimate eschaton,
from primordial creator to the last area and era, the last element,
species or stage of the known universe, if any or all of these
'lasts' are ever to be. A process that can be imagined to coagulate
in the course of its outgoing self, and to coagulate sufficiently to
form and allow to be separately identified the different stages,
species and individuals that we experience. And yet so that
none of these are so separate from the others that a genuine, unbroken existential linkage to all others, and especially to those
nearest in the actual process to any one individual or species,
time or place, cannot be discerned through all the border-signs
that may reasonably be erected between them.

If I take myself as an example of a quite distinctive individual, if I even pride myself, together with my fellow humans, on
enjoying the highest quotient of individuality to be found anywhere in the observable universe, I am yet very conscious of a
daily and nightly fact, and conscious of this fact since the dawn
of my individual self-awareness in the very womb of my mother;
the fact, namely, that this consciousness of my individuality is
achieved only and always in the context of inter-subjectivity.
That is to say, in the context of being conscious of other subjects;
in the context of a shared consciousness.[9] Similarly, if my
species, *homo sapiens*, is distinguished as an individual species

---

9. For an account of the so-called Intersubjective-First position in contemporary psychology, see *The Critique of Theological Reason*, pp 15ff.

from others by its rationality or intelligence, it is also by that very same token linked by an unbroken line to other species that betray intelligence, no matter how I define intelligence. And so on across all the kinds and species that are all so intimately inter-related and interdependent in this wonderful world.

A metaphysic that could give a reasonably verifiable account of our universe then, from originating source to ultimate end, would have to meet the following criteria. It would have to see mind and matter as co-ordinates in any perception of any part or range of reality. For we ourselves are essentially embodied minds, or intelligent bodies. And it is also true of any and every-thing we perceive in the observable universe, that they show as-pects of mind operating in and through the material universe. We do not directly perceive anything that is mindless matter or matterless mind. Although we may become aware of the pres-ence of both of these, but only through our observation of mind-formed-matter or matter-conditioned-mind. That is to say, we may become aware of matter or mind *per se* as limit cases. But in no case can matter or mind be seen as distinct substances. For mindless matter is what Aristotle called 'prime matter'; it is what might be termed the materiality of things, but not itself a thing or a kind of thing. Similarly, but differently for all we know, the matterless mind, or matterless mind-like entity, is one that is not circumscribed by or as any particular thing or things. And the difference resides in this: that it does not appear that matter in itself can ever exist as an entity on its own, whereas it appears that a mind-like entity, like consciousness might do so. Although it could not do so as a substance or thing as such, com-parable with and therefore separable from others. It would have to be *epekeina tes ousias*, as Plato put it, beyond being or sub-stance, beyond thingness that is of its nature finite. And the rea-son for believing that such can be true of mind-like entity, but not of materiality, is simply that mind-like entity seems more likely to be the source of material substance, rather than the other way round, from what we can see.

An example of the kind of metaphysics that should suit Susskind might be taken from Sartre, especially since Sartre took himself to be an atheist, or at least an agnostic, or in any case a metaphysician who was utterly indifferent to the god-question. 'Even if God did exist,' he has a character in one of his plays say, 'that would change nothing.' According to Sartre, there are two 'regions of reality': the For-itself or pure consciousness, that he calls the Absolute, and the formed and fashioned world, the In-itself, in which we live and work and die. 'And that's it! That's all', as Susskind might like to hear him say. The For-itself is a kind of no-thing-ness, for it is not in itself formed or restricted by any or indeed by all of the things in the world of which we are conscious. Indeed we come upon it indirectly. As we are conscious of any thing we are conscious of a consciousness that transcends any particular content of consciousness of any or all determinate things. That is how we become aware of absolute (not-thing-limited) consciousness. But the question arises from the point of view of this transcendent For-itself, how does it relate to the common universe of things of which our embodied minds are commonly conscious? Sartre answers in terms of the For-itself 'projecting possibilities' (no, he had not read Susskind), thus 'encircling a certain objective ensemble and throwing it into relief outlined on the world'. It is from these possibilities that human beings, in whom Sartre the existentialist is almost exclusively interested, and presumably other agencies operative in the universe, construct the realities that make for the coming-to-be of the universe that we know.

It has been observed by some of the more perceptive students of Sartre that his For-itself bears many resemblances to traditional Western notions of the One who is the undefined and indefinable source of all creation.[11] Although there is one important respect in which Sartre's account of the relationship between the For-itself and the empirical world reminds his readers more perhaps of the metaphysics of Eastern religious philosophy and imagery; as found, for example, in the Yoga Sutra,[12] than of the more creation-as-irresistible-fiat imagery of so much Western Christian metaphysics. And that in turn should remind us of some recent physicists who have turned to the East in search of metaphysical models that would prove helpful in their own trail-blazing efforts to go where modern physics seems to point, beyond the realm of direct observation that constitutes its most proven ground and the source of its most immediately verifiable and falsifiable insights.

---

11. See my *Christianity and Creation*, Continuum 2006, especially towards the end of the concluding Epilogue.
12. The Yoga Sutra is briefly summarised in *Christianity and Creation*, pp 183ff.

Finally then, two points to finish this postscript that has been devoted to the rather unscientific logic and assumption from which the contributions of scientists to the science-and-religion conversation so frequently suffer.

First, the finest example we have of a depth or height of reality that we can be aware of only indirectly through our awareness of forms of reality other than itself, occur at the innermost bourn or the furthermost limits of our very own consciousness. This is the self-conscious realm of consciousness. For self-conscious species like us – if there are other species like ourselves in this re-spect – in the mental act of knowing, or being aware of, of being conscious of other entities, are conscious of being conscious of these. That is to say, we are conscious of consciousness as such, operative in and through us. At this level we are not dealing with consciousness of any other formed and hence finite, limited entity. Hence Sartre, who also enumerates these steps towards this awareness of consciousness as such, terms it an absolute, and pictures it as a wholly other region of reality. Yet we only become conscious of this Absolute, ordinarily, in the process of being conscious of formed, finite things, and bending our con-sciousness back over and so beyond those content-full entities, in order to transcend our empirical consciousness of them, and to become simultaneously self-conscious and conscious of con-sciousness itself (a purely conscious Self?). 'Ordinarily' was in-serted into that last sentence because, as we well know, many techniques of meditation in Oriental religions are designed to empty our empirical consciousness of all its distracting contents in order to concentrate undisturbed on the pure awareness of pure Self. See the Yoga Sutra again, for example. And one should not try to prejudge the possible success of such efforts, even while in this condition of gross material bodiliness, in which our chances of divesting ourselves of all empirical aware-ness would seem to be remote.

Second, since therefore we cannot command a view of con-sciousness-as-such that might enable us to do this, we cannot entirely understand how this apparent essence of mind called consciousness can give rise to entities that are not pure con-sciousness, much less entities that do not appear to enjoy any consciousness of their own at all. Yet we know enough of this matter from our own experience, to be able to contradict Sartre's insistence that if a Pure Subject created anything, that thing could never escape from that Pure Subject's all-encompassing subjectivity, even so far as to be able to have an existence or life in its own right. Sartre, indeed, had an example of just such a possibility before his eyes and in his hands as he wrote *Being and Nothingness*, his metaphysical *chef d'oeuvre*. For the thought complex of that book, born of Sartre's mind, can be said to have taken on, quite literally, a life of its own ever since, through its incarnation in that book – a life that has already outlived J.-P., and may well live on after people have forgotten everything about him apart, perhaps, from his name.

But we have here in this example, not just an *argumentum ad hominem* against one of Sartre's least credible observations on the subject of the Absolute. We also have a metaphor or an analogy for the manner in which Pure Consciousness or Mind or Self can create a material world, in which evidences of mind are present in varying degrees. And since we are ourselves beings who are duck-rabbit-like bodily-embedded minds, who can see directly only mind equally embedded to varying degrees in all parts of the physical universe, metaphor and analogy is all we can hope for, unless and until some deeper transformation is to take place in our current mode of existence. Now this metaphor or analogy then tells us, incidentally, that it is from mind that what we call matter emerges as, to continue the metaphor, its thoughts take shape and take time – thus yielding the four dimensions most familiar to us that make up our familiar world of space-time. Then what is meant by matter will finally turn out to be something like Susskind's 'landscape of possibilities', with the land imagery in that phrase being close enough to Aristotle's idea of 'prime matter' as the primordial realm of all potential emergences. At the very least, this view of the matter might save us from the further tedium of outed or closet materialists of the crude dualist kind whining on and on about the difficulty of explaining how consciousness emerged from this matter, from which it was, on the basis of their crude dualist materialist dogma, originally and by definition declared absent.

If these concluding remarks sound peremptory, it is because they are; and because a lengthy treatment of the crucial topics they contain would be out of place in what is simply a kind of tidying-up epilogue to a collection of essays. So that the very least that can be said of them is this: they emerge so naturally from an analysis of the conceptual clarity, or lack of it, and of the logical integrity, or lack of it, shown by some scientists in the course of conversations with theologians, that they deserve the most serious consideration by any scientists who do decide, as is their native right, to infiltrate the porous border that never can altogether divide physics from metaphysics. Indeed, if one may be so bold as to say so, in agreement with Schrodinger,[13] any scientist who wishes to make the crossing could well benefit in terms of both conceptual and logical proficiency from a visit to the mighty metaphysician-theologians of the early Greek past and down to Hegel at least, who may have known as little about the physical sciences as these much feted modern scientists know about metaphysics.

Should there be a Concluding Untheological Postscript?
Yes, but not here. Apart from the fact that parts of it are contained in the preceding chapters, the whole of such a critique of the kind of false theological positions that pose persistent and serious problems for the science and religion dialogue, are dealt with elsewhere in the length they require, if only because many of these pose equal problems for dialogue with anyone at all who lives outside the faith communities in which such positions are formulated.[14] So it is necessary here only to record briefly the positions of the principal enemies of received scientific wisdom, who act in the name of religion. And all of these here are from the Christian tradition.

First, and most generally, the fideists. These maintain that the truths of the Christian faith, including those that concern the creation or goal of our universe, are beyond the range of human reason, either because human reason is too weak to penetrate such matters, or because humanity has been so corrupted by the sin of Adam that it can no longer manage to master them. People must therefore accept all the truths of the Christian religion on faith alone, blind faith as their interlocutors would call it, in a God who especially revealed these truths to the prophets of the Israelites and, definitively, to and through the prophet, Jesus of Nazareth. That this is a false view of the matter is evident both

---

13. See Erwin Schroedinger, *Nature and the Greeks*, Cambridge University Press 1996, ch 1.
14. For instance in *The Critique of Theological Reason*, especially the section on 'Beginnings Old and New;' and in *Christianity and Creation*, especially in the Epilogue.

from the said prophets and from the teaching and life-work of the historical Jesus himself, none of whom ever deny that God the creator can, and can at all times, be known from our experience of the creation itself.

Second, and as a sub-set of the former group, the biblical fundamentalists base themselves, not really on the Bible as it is and as we have it today, but on a dogma about the Bible that is of their own invention, and on a profound misunderstanding of the nature of the biblical literature. The dogma that they bring to the Bible, rather than find through reading it, is the dogma of the literal inerrancy of the Bible in all of its pages. The misunderstanding of biblical literature is manifold, but for present purposes it consists largely in the failure to see both the virtual omnipresence of myth throughout the Bible, and the nature of myth as a fundamental means of imaging and expressing the most fundamental truths about the cosmos and about all that are active in it. String theory is myth in so far as it deploys its story of the emerging universe by imagining string-like entities vibrating, or imagining waves, fields, attractions, repulsions, and so on. This misunderstanding culminates in mistaking a mythic story of the creation of the world, for instance, for a piece of plain history, which is to be received quite literally as plain history – rather than be received as myth must be, quite literally as metaphor.

The main outcomes of the combination of this dogma with this misunderstanding, or at least the outcomes that mainly affect in adverse fashion the necessary dialogue with the scientists, are as follows: first, the concept of divine creation as a one-off event that happened a finite length of time ago. The more brazen of these fundamentalists would set that time in terms of thousands of years. Second, the image of the act of creation as that of a peremptory and irresistible command of the kind: 'Let there be ... and there was.' Third and consequently, the insistence that the creation thereafter is much as it was then, and God does not need to intervene again, except to prevent certain creatures who were created with free will from wrecking the place and themselves, in their wilful abuse of that prerogative. Otherwise, that is to say, if human sin had not intervened, God could have stayed in heaven, let the thing run as it had been set up to run, and receive human beings into heaven after they had lived their lives as they were supposed to have lived them, without sin. (That, in any case, is a fair summary of Calvin's account of the matter in his *Institutes*.)

When the biblical account of creation is read properly, as truth discovered and expressed in mythic form, that is to say, through image, metaphor and story, the act of divine creation is not a one-off act of raw divine power, but a continuous act of forming and transforming. Forming things by formulae, in forms that can then create new forms ('let the earth bring forth'), the emergence of novelty, and ultimately evolution.[15] And these dynamic forms are processes of mind, as Plato understood in all his talk of the forms. So that mind- or consciousness-like entity and not matter is, to use Schroedinger's phrase, the fundamental concept. Mind-like entity forever creating evolving universe(s) from within; being embedded in material substance, or material substance embedded in it: the double immanence of the biblical myth of creation. God in creatures; creatures living and moving and having their being in God. So that as far as we self-conscious creatures can see directly, as deeply within ourselves and as far outside ourselves as we can see, there is this reality plotted always according to the co-ordinates of mind and matter, or of subject and object. As Schroedinger also put it: 'The world is always given to me only once, not one existing and one perceived. Subject and object are only one. The barrier between them cannot be said to have broken down as a result of recent experience in the physical sciences (quantum physics, that is), for the barrier does not exist.' Although of course we can detect through and only through these mind-embedded but matter-limited substances both an all-originating mind-like entity that is unbounded, unformed fullness in one direction; and an ultimate limit-entity that is the unbounded, hence unformed emptiness in the other direction. This is the nothingness that comes into existence at the limits of finite being, as darkness comes into existence at the outer limits of a ray of light. The *ne plus ultra*, like the nothingness that surrounds the spherical, if these are spherical spatial boundaries of the universe; a limit case of the materiality of things that Aristotle called 'prime matter.'

In *Black Holes and Baby Universes* Stephen Hawking tells us 'that both time and space are finite in extent, but they don't have any boundary or edge. They would be like the surface of the

---

15. Erwin Schroedinger, *Science and Humanism*, Cambridge University Press 1966, p 122.

earth, but with two more dimensions ... If this proposal is correct, there would be no singularities, and the laws of science would hold everywhere, including at the beginning of the universe. The way the universe began would be determined by the laws of science. I would have succeeded in my ambition to discover how the universe began. But I still don't know why it began.'

A cosmic singularity here refers to a stage or state of the production of the universe prior, if only logically, or an infinitesimally brief period of time, prior to the laws of science taking over to determine all of its further coming to be. And the rejection of such singularity would remove the idea of a creator divinity who, by irresistible *fiat*, first put the stuff there out of which 'the laws of science' could determine the rest of the universe's coming-to-be. So Hawking, by rejecting such singularity and its producer, rightly as it happens, believes he can then say he explains by the sole operation of the laws of physics, how the universe came and comes about. But he then acknowledges that he will still not know why the universe emerges from beginning to end, for he will not know who or what sets these laws to work, or for what purpose this is done. Is this a possible prop, graciously extended to a religious worldview built around a creator God, but constantly assaulted in the name of science? If so, it is of real use to the religious worldview, and that for two reasons. First, because Genesis describes the beginning of the universe in terms of forming, not in terms of omnipotent *fiat*s. And such forming is from beginning to end effected by formulae comprehended under 'the laws of science'. And God is simply the name for the creative source of the universe whose formulaic laws are indicative of the existence of mind, and who therefore sets these laws in action.

Second then, Genesis also, and the whole Bible after that, gives the correct answer to the other part of Hawking's why? The part that asks, for what purpose were these creative formulae set in action so that the evolving universe should continue to emerge? The universe is created because, as the Genesis story repeats like the flashing of a neon-sign, it is good, it is good, it is very good. Existence and life and all of its transformations to higher and higher status and quality is good for those who enjoy rather than attempt to destroy it. And that is quite enough, and more than enough *raison d'être* for all but the most inveterate of human begrudgers, to be getting on with. But that is also an answer that can be read off from all that happens in the universe as a result of all the agencies that are operative within it: plainly, the simple and interior purpose and aim of all for all is the maintenance and advance of existence and life and its greater enjoyment, up to no discernible limit.